ENCOUNTER SERIES

Guaranteeing the Good Life: Medicine and the Return of Eugenics

Essays by

Hadley Arkes
Brigitte Berger
James Tunstead Burtchaell, C.S.C.
Robert A. Destro
Jean Bethke Elshtain
Stanley Hauerwas
Christopher Lasch
Richard John Neuhaus
Paul C. Vitz

and
The Stories of Two Encounters by
Richard G. Hutcheson, Jr.

Edited and with a Foreword by
Richard John Neuhaus

WILLIAM B. EERDMANS PUBLISHING COMPANY
GRAND RAPIDS, MICHIGAN

Published by Wm. B. Eerdmans Publishing Co.
in cooperation with
The Rockford Institute Center on Religion & Society

Library of Congress Cataloging-in-Publication Data

Guaranteeing the good life: medicine and the return of eugenics /
    essays by Hadley Arkes . . . [et al.]; and the stories of two encounters
    by Richard G. Hutcheson, Jr.; edited and with a foreword by Richard
    John Neuhaus.
        p.   cm. — (Encounter series; 13)
    ISBN 0-8028-0213-3
    1. Medical ethics. 2. Eugenics. I. Arkes, Hadley. II. Neuhaus,
Richard John. III. Series: Encounter series (Grand Rapids, Mich.); 13.
R724.5.G83   1990
174'.2—dc20   89-25922
                                                         CIP

Contents

Part II: The May Conference

Foreword

It may seem that there is nought, or at least very little, for our comfort in the following pages. The subtitle is *Medicine and the Return of Eugenics*. But the book is about much more than medicine, for medicine is about much more than is usually meant by the term "medicine." It is about anthropology—what and who we think human beings are. It is about moral ends—what we mean by health and wholeness, and the other "goods" to which these ends are related. It should not be surprising, therefore, that the essays and the discussions in this volume take up questions of law, history, psychology, cultural criticism, and much else.

But the subject keeps coming back to what is being done in hospitals, nursing homes, and laboratories—and to the questions about what people are proposing to do in such places. Some of what is discussed is intriguing and hopeful; some of it is troubling and even terrifying. But note that, while it does look also to the future, this book is not an exercise in "futurism." The developments treated are not science fiction. These are the things that are being done now, and the things that are being planned, usually for the immediate future. All the participants in this book are keenly aware of the importance of not being alarmist. Developments are alarming enough without alarmism. The reader will recognize in these pages, I trust, an unremitting and sober sense of urgency. Many years ago now, C. S. Lewis wrote, persuasively and chillingly, about "the abolition of man." He was not referring to the obliteration of the human race by nuclear war or some other catastrophe. He meant, rather, the technological redesign of human nature in such a way that "human nature" would be lost. That might strike many writers as a fantasy. Yet it is a fantasy brought closer by what is now being done and, more important, by the reasons given for the doing of it.

The first essay in the volume, which I wrote, was not a paper prepared for the "Good Life" conference. It serves as an introduction to the concerns that gave rise to the conference. Following that piece are the conference papers. Hadley Arkes, the distinguished moral philosopher at Amherst, explores connections between technology and culture, whether we do not lose control as human beings when we attempt to assume control of the human condition. Cultural historian Christopher Lasch examines relationships of "natural" obligation, such as those of the family, suggesting that the only progress we should welcome is a progress firmly rooted in cultural conservatism. Jean Elshtain, the political scientist now at Vanderbilt, analyzes the various "feminisms" that have had such a powerful impact on our understanding of "the good life" and issues raised by, for example, medical technology. Father James Burtchaell of Notre Dame examines those institutions, such as the Roman Catholic Church, that try to provide a place to stand in articulating a critical assessment of otherwise threatening changes. Sociologist Brigitte Berger of Boston University helps us to locate the social contexts in which distinctive ideas of the good life are manufactured, marketed, and maintained. Stanley Hauerwas of Duke University dissects the moral and religious difficulties we have in accepting those human circumstances that offend our ideas of the good life. Robert Destro, professor of law at Catholic University, argues that against the seemingly inexorable forces of technology there may be some potential for resistance in law, especially in civil-rights law. But, as he makes clear, law has its own seemingly inexorable forces. Finally, Paul Vitz, professor of psychology at New York University, takes us through the changing assumptions of his discipline, and the way those assumptions may be leading to a new apprehension of what makes life good.

Well, not quite "finally." Finally is the two-part "Story of an Encounter," written by Richard Hutcheson, who at the time of composition was a senior fellow of The Center on Religion and Society. As with other books in the Encounter Series, the reader will discover here some of the liveliest engagements with the questions posed. There is sharp disagreement among the participants, and frequently a shared sense of the impossibility of getting a handle on lines of thinking and doing that seem to be fast spinning out of control. As said at the outset, it seems there is nought, or at least very little, here for the reader's comfort.

It is hoped that most readers will not take up this book in the expectation of being comforted. The purpose is to inform, to stir thoughtful concern, and—just maybe—to indicate alternatives to the ominous future that is already upon us. We have not the right to despair. And, if the final words of hope with which the book concludes seem disturbingly religious to some, what has gone before those final words should explain why that must be so.

I am pleased to acknowledge my debt, once more, to my colleagues Paul Stallsworth and Davida Goldman, who make these explorations happen.

NEW YORK CITY   Richard John Neuhaus

The Return of Eugenics[1]

Richard John Neuhaus

Eugenics—that is, the movement to improve and even perfect the human species by technological means—arose in the late nineteenth century and flourished in this country and in Europe until the 1930s. Then it was challenged by scientific counter-evidence, and by growing uneasiness about its racialist implications. Later, or so the story was told, eugenics was definitively discredited by the Third Reich, which enlisted its doctrines and practices in support of unspeakable crimes against humanity. But now, in the journals and in the textbooks, the story is being told differently. The problem, it is said, was not so much with eugenics itself but with the Nazis: they abused eugenics, they went too far, they were extremists.

Thus, in the longer view of history, the horror of the Third Reich may have effected but a momentary pause in the theory and practice of eugenics. For today, four decades later, eugenics is back, and it gives every appearance of returning with a vengeance in the form of developments ranging from the adventuresome to the bizarre to the ghoulish: the manufacture of synthetic children, the fabrication of families, artificial sex, and new ways of using and terminating undesired human life.

To be sure, the literature on all sides of the current disputes about these developments remains riddled with references to the

1. Reprinted from *Commentary*, April 1988, by permission; all rights reserved.

1

Nazi experience. But the mention of troubling similarities to the Third Reich is, as it should be, accompanied and qualified by other observations. No responsible parties suggest that America is, or is likely to become, Nazi Germany. That is patently absurd. What happens here is and will be distinctively American. And, because this is America, there are political, legal, and moral resources to resist scenarios of the worse inevitably coming to the worst.

In addition, the great majority of today's eugenists take pains to distance themselves from any hint of racialism, although some very respectable proponents of "population control" are not averse to writing about "inferior" population groups. Further, it must be acknowledged that there have in fact been very impressive technological advances, some of which are indeed breakthroughs to uncharted regions of control over the human condition—and some of which hold high promise for reducing misery and enhancing life. There is no room in this discussion for Luddite reactionaries who claim to discern in every technological change the visage of the "brave new world." Finally, those who take a favorable view of the developments in dispute would seem to be motivated by the best of intentions. With few exceptions, their language is the winsome one of progress, of reason, and, above all, of compassion.

All that said, we *are* nonetheless witnessing the return of eugenics. And with it have come questions, inescapably moral questions, for which we appear to have no good answers. Indeed, it is doubtful that we still have a sufficiently shared moral vocabulary even for debating what good answers might be. Yet whether we like it or not, these questions are already being answered from one end of the life cycle to the other in terms that, were one not wary of alarmism, might be described as alarming.

To begin, literally at the beginning, consider first the putting together of human gametes (sperm or egg) in order to facilitate new ways of having babies, and to produce babies of higher quality.

It is hard not to sympathize with couples who want children but cannot have them because of infertility. What are they to do? Of course there is adoption, but the "right kind" of child is hard to find, and getting one can be very expensive. (One and a half million abortions per year have put a considerable dent in the supply side of the American adoption market, plus adoption is one area where discrimination on the basis of race or handicap is still eminently respectable.) In addition, many people want a child that is, at least in part, produced from their own biological raw materials. Techniques for meeting this market demand are several.

Artificial insemination by husband (AIH) has been with us for some time and is relatively straightforward. Artificial insemination by donor (AID) is technically identical but introduces a third party to the relationship, or, more accurately, biologically excludes the husband. These techniques are really quite simple and lend themselves to do-it-yourself procedures—with a little help from your friends.

A recently publicized example was the case of an Episcopalian priest who wanted a baby but definitely not a husband. She invited three friends over (two of them priests) to masturbate for her, and she then impregnated herself with the mixture of their sperm. The purpose of having several sperm sources, she explained on national television, was to avoid knowing who the father was, and thus to make sure that the child would have an intimate bond to no one but herself. The child is now three years old, and the mother has declared that she intends to have another baby by the same procedure. The *Washington Post* described her as the first artificially inseminated priest in history, which is probably true. Her bishop, Paul Moore of New York, appeared with her on television and gave his unqualified blessing to this undertaking, citing the need for the church to come to terms with the modern world.

In vitro fertilization (IVF) is yet another procedure. It involves the woman being given hormones to stimulate egg production. The eggs are then "harvested" and mixed with sperm from the husband or someone else, and some eggs that become fertilized are placed in the woman's uterus. But ethical questions have been raised about the use and disposal of the many "superfluous" embryos that are not transferred to the uterus. Some practitioners of this technique—confronted with the argument that anything with the potential of becoming human life is human life—resolve the problem to their moral satisfaction by declaring that very little embryos are "pre-embryos." On the other hand, there is intense interest—on the part of drug companies and genetic researchers—in letting such embryos develop in the laboratory so that they can be used for scientific experiments.

An additional problem is "superfluous" fetuses. Because the technique is time-consuming and expensive, and because success is by no means assured, "extra" embryos are placed in the uterus in the hope that at least one will "take." With disturbing frequency, this results in two or three or more very healthy fetuses. When the procedure produces more fetuses than the mother wants or, in some cases, than she can safely carry to term, the practice is to use an ultrasound probe to guide a needle which punctures the hearts of the fetuses to be eliminated. Doctors who do this allow that there may be a moral problem in terminating fetuses that they helped bring into being. But then, it is observed, the morality of the thing is really not that different from the elimination or experimental use of unwanted embryos.

Yet another option for the infertile couple is to elect embryo transfer. Here the husband, or someone else, contributes the sperm with which a "donor" woman is inseminated. The resulting embryo is then "washed" from her uterus and placed in the infertile wife. This procedure is still somewhat experimental and poses high risks to both the donor woman and the embryo, but we are assured that progress is being made in ironing out the difficulties.

An additional option, surrogate motherhood, was the center of a media storm in 1986 in connection with the Baby M case in New Jersey. The terminology is misleading, since the woman is not a surrogate or substitute but is in fact the mother. The procedure might better be called "term" or "contract" motherhood, for she contracts to act as mother only until the child is brought to term. More than a dozen states are now considering measures to legitimate contracts for such rent-a-womb arrangements, but the debate over the practice has deeply divided feminists, and the Left generally. The anxiety is over the using of women, typically women who are vulnerable and in need of the money. There is also objection to what is, after all, a particularly gross capitalist act, even if between consenting adults.

Writing in the *Nation*, Katha Pollitt further complains that the man in these cases wants "a perfect baby with his genes and a medically vetted mother who would get out of his life forever immediately after giving birth." (Some contracts also stipulate that the mother will abort the child if there is evidence that it is not up to standard.) Miss Pollitt observes that no other class of father—natural, step, or adoptive—can lay down such conditions. While making some predictably contemptuous remarks about the Vatican's position on reproduction, she does side with Rome about one thing. "You don't have a right to a child, any more than you have a right to a spouse. You only have the right to try to have one. Goods can be distributed according to ability to pay or need. People can't. It's really that simple."

But Katha Pollitt and others of like mind do not appreciate the reach of the eugenic vision, which is to eliminate the limits and risks in what was once deemed to be natural. In any event, contract motherhood is but a very small part of the transformations now under way. In ten years the procurers in that business have been able to sign up only five hundred women. It is an enterprise that fades in comparison with the real growth areas in the synthetic-child business.

In the past, a distinction was drawn between positive and negative eugenics (though in usage the terms were sometimes reversed). Positive eugenics was thought to be relatively innocent, simply a matter of breeding good human stock in order to improve the race by increasing the number of physically and mentally fit. Negative eugenics, on the other hand, made a lot of people nervous, since it meant preventing the birth of the unfit or eliminating them after they were born. In America in the 1920s more than half the states had mandatory sterilization laws applicable to people who fell into various categories of unfitness. (The laws were enforced mainly in California.) And of course, as the textbooks say, Nazi Germany took both positive and negative eugenics altogether too far.

The distinction between positive and negative eugenics is no longer always helpful. For instance, intervention to eliminate a defective gene rather than to eliminate a defective fetus may be viewed as either positive or negative. Still, some of the more striking changes today are in the area of the positive improvement of the human stock. Indeed, what is now being done and proposed makes earlier efforts at improving the race (for example, the socially and morally clumsy *Lebensborn* program for breeding the SS elite with superior Reich female stock) seem pitiably primitive.

At present, research focuses on detecting and remedying genetic ills or ailments by removing or adding genes. But discontents with the human condition as it is now constituted are almost infinitely expansive, and since it is almost impossible to argue against the proposition that the quality of human beings we have been turning out to date leaves much to be desired, the pressure to move the limits of intervention may be nearly irresistible. Is asthma a genetic disease? If asthma, why not baldness, or shortness, or having the "wrong" color eyes? And surely we still focus on diseases only because we have this ancient idea that medical interventions should be therapeutic. Instead of restricting ourselves to curing diseases, however broadly defined, why not be more positive and aim at the desiderata of human life? The combination of reproduction technology and engineering, or either one by itself, may be able to assure the production of socially desired personality types. In that event, presumably, "society" will decide which types are desired.

The enzymes that slice DNA produce nucleotide bases that scientists call "sticky ends" because they merge so easily with the genetic structure of another organism. Many are troubled precisely by the sticky ends to which this technology is being put. But there are some who are not troubled.

Lloyd McAuley, a New York patent lawyer, has written, "I understand the fear that we may be letting the genie out of the bottle as we expand our ability to alter biological evolution. I do not share that fear." He allows that there should be some control over developments "until we learn a bit more about where we are going." But, all in all, there is no cause for anxiety since what is happening is not really so new. "Switching genes around strikes me as little more than expedited breeding," he writes. "As an ethical issue, whether or not we wish to do that with human beings may not be much different from whether or not we wish to breed human beings." "Expedited breeding"—it is a reassuring phrase.

Another comforting voice over the years has been the editorial page of the *New York Times*, although, to be sure, the comfort is attended by stylistic rumblings of deeply pondered concern. Whether the subject is genetic engineering or experimenting with human embryos, the *Times* typically informs us that it is too late to raise the kinds of questions we want to ask. For example: "Critics are concerned that making life forms patentable will give animal and eventually human life too much in common with commodities, leading to disrespect for both. But society has already passed that point." On the *Times* editorial page, the big decisions are made by society, and society is forever busily bustling along. The editors simply report their sightings of it as it passes one point after another.

The *Times* acknowledges that genetic engineering is "at first sight disquieting," but the editorialist has taken a second look and concludes that "It's hard to object to improving a species' inherent characteristics." As to problems we may have with engineering that does change the "inherent characteristics" of a species, we are told that "Such conundrums still lie in the realm of science fiction." They may be as much as ten or twenty years off, and, as John Maynard Keynes suggested, in the long run we are all dead.

One cannot help being struck by the blithe assumption that we can still agree on "the inherent characteristics of a species" — of the human species, for instance. For we have, after all, been through a systematic assault upon the idea that there is anything "inherent" or "natural" in the makeup or behavior of human beings. With respect to sexual identity and behavior, gender relations, familial bondings, and a host of other questions, the human condition is declared to be boundlessly various and malleable. In all these areas, the protester's appeals to what is natural are dismissed with enlightened contempt. But now, as we intervene to restructure human beings genetically, technicians and their apologists assume the tone of Thomistic philosophers explaining the self-evident truths of natural law, assuring us that they recognize and will respect what is natural, inherent, and essential to being human.

This is, at best, an instance of what Allan Bloom calls debonair nihilism. More likely, it is a desperate effort to conceal from others, and from themselves, the consequences of what they do; of what they cannot bring themselves not to do—because it is possible, because it is progress, because the adventure of doing the thing that could never before be done is near to irresistible.

One such thing is the use of fetal tissue. Fetal brain, pancreas, and live tissue is, it is said, admirably suited for the treatment of Parkinson's disease, Alzheimer's disease, Huntington's chorea, spinal-cord injuries, and leukemia. Fetal tissue is also excellent for implant treatments because it grows faster, is more adaptable, and causes less immunological rejection than adult tissue. Whatever one thinks of abortion, it is argued, it is a shame to let the material go to waste. There are literally millions of people who might benefit from these human parts. Dr. Abraham Lieberman of New York University Medical Center says of these developments, "This is to medicine what superconductivity is to physics."

Admittedly, there are concerns about collusion between abortionists and physicians, about how to decide whether the fetus is actually dead, about commercial trafficking in fetal parts, and about women becoming pregnant in order to produce fetal parts to order. Those are only some of the concerns that have been raised. But the decision to move ahead on this front is, we are told, another point that society has passed. As the director of the American Parkinson Association observes, "The majority of people with the disease couldn't care less about the ethical questions—they just want something that works."

Both pro- and anti-abortion groups have expressed uneasiness about the use of fetal parts. Pro-choice groups worry that, as with contract motherhood, it could invite the exploitation of women's bodies to produce custom parts, as it were. Pro-life groups worry that it could make abortion seem more attractive to some women because the parts would be used to help other people. "The worst possible ethical evil of all this," says Arthur Caplan of the Center for Biomedical Ethics at the University of Minnesota, "would be to create lives simply to end them and take the parts."

Unencumbered by the delicately nuanced inhibitions of some ethicists, however, the general media are generally enthusiastic. *Newsweek*, for instance, allows that some controls are necessary "to keep fetal research from becoming barbarous." But that does not blunt *Newsweek*'s keenness on the new technology which "has created a surge of interest in fetal-tissue implantation, and research both here and abroad is beginning to offer an exciting glimpse at treatments that could lie ahead."

Recently a California woman asked a medical ethicist whether she could be artificially inseminated with sperm from her father, who has Alzheimer's disease. She intended to abort the fetus so that the brain tissue could be transplanted into her father's brain. The usual response to such questions is that this entire field is still in its infancy, so to speak, and clearer guidelines are yet to be developed. But the California woman's act of love for her father would no doubt meet with overwhelming support on the Phil Donahue show and similar popular seminars in contemporary ethics.

To more thoughtful students of these matters, the use of dead fetuses leads to some surprising confusions. Britain's Warnock Committee, for example, recommends that there be a 14-day-cutoff rule for experimentation on embryos that are fertilized in the laboratory. Charles Krauthammer, writing in the *New Republic*, basically agrees, while acknowledging that the 14-day rule may prepare society for 14 weeks or 40 weeks. "Does any such rule not place us on a slippery slope?" he asks. "The answer is that society already lives there. In fact, it has slid far beyond the 14-day period. In most English-speaking jurisdictions, one can do with an aborted fetus that is many weeks old pretty much what one wants: discard, research, implant. The 14-day rule moves us further up the slope from where we are today." Krauthammer is surely right about the slippery slope. (I am sometimes asked whether I "believe in" the slippery slope, as though it requires an act of faith. I believe in the slippery slope the same way I believe in the Hudson River. It's there. There is no better metaphor to describe those cultural and technological skid marks which are evident to all who have eyes to see.)

The oddity of the Warnock recommendation, however, is that its concern for the dignity of human life results in greater respect being shown for those who are fertilized in vitro (in the laboratory dish) than for those fertilized in vivo (in the body). Krauthammer suggests, delicately, that we cannot think clearly about the new questions related to the production and exploitation of human life without rethinking the old question of abortion. Because they view the abortion debate as wearied and wearying, as polarized and stalemated, many will resist that suggestion.

In any event, it can be argued that the eugenics project—in both what is proposed and what is already being done—has moved beyond disputes about life before birth, or about life that was never intended to be born. Once again society has passed that point. The new and more "interesting" questions have to do with the termination and medical exploitation of human beings already born. In September 1987, the *Newsweek* story was titled "Should Medicine Use the Unborn?" Having answered that question affirmatively, in December (three months later) the question agitating the media was "Should Medicine Use the Born?" The opening wedge to this new phase was the debate over what might be done with anencephalic newborns (babies born with most of their brains missing).

As with almost all the questions considered here, New Jersey is vying, successfully, for the honor of being in the legal vanguard. Assembly Bill 3367 would permit parents to donate the organs of an anencephalic child. At present, they have to wait for the child to die first. The new law removes that technicality by declaring the baby dead before it dies. In California, however—confident that the law would quickly catch up with practice, and ethics with the law—they did not wait for a change in law.

Loma Linda University Medical Center is connected with the Seventh Day Adventists, a highly moral, even moralistic, religious group that insists on abstinences from alcohol, tobacco, and—although not universally—tea and coffee. Everything done at the medical center is subjected to the strictest ethical scrutiny. The center, like hospitals most everywhere these days, has a highly qualified ethics committee. Indeed, it has been observed, correctly, that in the last two decades medical technology has been the salvation of ethics as a profession. Thousands of medical ethicists and bio-ethicists, as they are called, professionally guide the unthinkable on its passage through the debatable on its way to becoming the justifiable until it is finally established as the unexceptionable. Those who pause too long to ponder troubling questions along the way are likely to be told that "the profession has already passed that point." In truth, the profession is usually huffing and puffing to catch up with what is already being done without its moral blessing.

The star of Loma Linda is a surgeon named Leonard Bailey. On October 16, 1987, Dr. Bailey led a team that transplanted a heart from an anencephalic infant into another baby delivered by cesarean section three hours earlier. A statement from Loma Linda notes that the procedure was "innovative medically" and also "interesting ethically" because it "prompted further discussions regarding the moral wisdom of using brain dead or non-brain dead anencephalic human neonates as organ donors." ("Neonate" is the term for children less than a month old.) The girl baby who "donated" her organs was, interestingly enough, named Gabrielle, the feminine of Gabriel, the archangel who reveals the things to come.

There are about 3,500 anencephalic children born each year in this country, and most of them die within a month. The problem is that their organs deteriorate and are not much good for transplanting if you wait until they die. Loma Linda recommends to parents that the children be allowed to live for no more than a week before taking the parts. The parents, we are told, find this procedure "deeply meaningful," since their disappointment in having a handicapped baby is "redeemed" by putting the baby to good use in "helping others." The language of redemptive suffering is very prominent in the discussion of these matters. The sorrow of being afflicted with handicapped children or older people with severe disabilities, we are informed, is significantly assuaged by "donating" them for altruistic purposes.

Dr. Jacquelyn Bamman, a neonatologist, is among those who are troubled about what is now being done and proposed, well knowing that today's somewhat speculative proposal may be next week's *fait accompli*. Dr. Bamman worries about the clear departure from traditional medical ethics by doing surgery that is not intended to benefit the child, and indeed is directly aimed at causing the death of the child by removing the heart. She notes "the lack of any rational way to prevent the extension of this same approach to involve other children with serious defects." If the prospect of a limited life-span justifies the killing of children in order to use their organs, the issue goes far beyond the anencephalic to include children with Tay-Sachs, Werdnig-Hoffman, and other diseases. It might be argued that in these cases, unlike the anencephalic, there would be "benefit" to the children since it would relieve them of pain (it being assumed, although no one can know for sure, that the anencephalic feel no pain). If all whose brains are severely abnormal are potential "donors," Dr. Bamman observes, the field is opened to infants with hydranencephaly, grade IV intracranial hemorrhage, Trisomy 13 and 18, and a host of other handicaps.

Responding to Dr. Bamman on behalf of the Loma Linda

ethics committee, Dr. O. Ward Swarner acknowledges that she has indeed raised some interesting questions. He assures her that at the present time "there are no intentions or justifications for putting some in jeopardy to harvest organs for others." At the same time, the ethics committee is in constant "consultation with other concerned staff members, nurses, social workers, ministers, and ethicists" and will "follow with interest" the work of other experts in this rapidly developing field. Dr. Swarner firmly states that "the ethics committee has not approved any harvesting of organs or procurement of transplants in any other than brain-dead patients."

But, of course, tomorrow is another day. Speaking of a mother who agreed to have her baby's organs harvested for others, Dr. Joyce Peabody, chief of neonatology at Loma Linda, said, "She has made a major contribution by getting us brave enough to face this issue head on." We can be confident that the brave surgeons and ethicists of Loma Linda will not flinch in the face of the next "technological breakthrough." Nor is it likely that other institutions will long allow the stars of a few institutions such as Loma Linda to dominate the firmament of the bright new world now in sight.

To be sure, there are those who warn against the seductive appearance of the brave and the bright. The late Paul Ramsey is sometimes called the father of contemporary medical ethics, and he had reason to rue much of what he helped to wreak. Testifying before a government committee on medical ethics three years ago, Ramsey said, "I respectfully express the hope that the committee will be initially prepared to say 'Never' to a number of things that are now being done or proposed and that are now proximately possible to be done, and not merely to things that may be only remotely possible. Remote possibilities are soon proximate, and soon done."

But what about just saying no? It is possible to say it, but much more difficult to make it stick. Even a good reason for saying no makes little impression in a culture that has lost any shared understanding of the good. Pitted against every "no" is the logic of progress, the ambition of pioneers, and, not to put too fine a point on it, the lust for fame and fortune. Even those who have the nerve to say no almost never say never. Then too, and also hard to resist, is the impulse of compassion—to relieve the suffering of "meaningless" human lives, to contribute to the health and happiness of others.

Actually, organ transplants involving infants are still highly experimental. As of this writing, there have been only nine heart transplants involving newborns, four kidney transplants, and no liver transplants. But technology proceeds apace, and those who say no—never mind never—are politely but firmly informed that medical practice has already passed that point. And, of course, we are not talking only about infants, although for some reason "breakthroughs" in what we give ourselves permission to do to people usually begin with little people, and with the old or very sick.

There is, at all stages of life, an obvious connection between the harvesting of healthy organs and the decision about when someone is dead, or should die. The question of euthanasia is thus an integral part of the progress of the eugenics project.

Of course, the dispute over the merits and demerits of euthanasia has been with us for a very long time, going back to the Greeks and Romans, long before people attributed their decisions to the force of technological breakthroughs. But today the discussion is taking interesting turns.

The Dutch, it is generally acknowledged, are a very progressive people. That country's program of voluntary euthanasia, which is said to account for up to 8,000 deaths per year, has recently received a great deal of attention in the American media. In the last year several television programs have dramatically contrasted American practice with the more advanced and humane approach of the Dutch. A report in the *Wall Street Journal* declares, "The Netherlands is pioneering in an area that in the coming decade is likely to be a focus of medical, legal, ethical—and intensely emotional—debate in many industrialized countries." A spokesman for the Royal Dutch Medical Association explains, "What we are seeing now is the result of processes and technology that keep people alive too long, people who are suffering, people you cannot help in any real way." Daniel Callahan of the Hastings Center has made an intriguing contribution to our language by describing such people as the "biologically tenacious."

Not everyone, it should be noted, is enamored of the Dutch program. For instance, Dr. Richard Fenigsen of the Willem-

Alexander Hospital in the Netherlands cites a number of studies indicating that one problem with the voluntary-euthanasia program is that it is frequently not voluntary. At some of the major hospitals, general practitioners seeking to admit elderly patients are advised to administer lethal injections instead. Involuntary active euthanasia (direct intervention to terminate a patient's life without the patient's permission) has not yet been incorporated into law, but there is great judicial leniency. For example, a doctor suspected of killing twenty residents of a senior-citizens' home pleaded guilty to killing five, was convicted of killing three, and was given a fine.

As might be expected, these developments both reflect and effect changes in popular attitudes. In a recent Dutch opinion poll, 43 percent of the respondents favored involuntary euthanasia for "unconscious persons with little chance of recovery." On another question, 33 percent had "much understanding" and 44 percent had "some understanding" for those who, out of mercy, kill their parents without their consent. Seventeen percent thought it "probable" that they would ask for involuntary active euthanasia for a demented relative.

The synod of the Reformed Church in Holland, desiring to offer moral guidance on coming to terms with the modern world, is perceived to be quite favorable in its attitude toward involuntary active euthanasia. Dutch of less advanced opinion, on the other hand, claim to have noticed a striking upsurge in the suspicion expressed by the elderly and sick toward doctors, hospitals, and their own families. (A Gallup poll reports that four times as many Americans would donate a relative's organs as would donate their own. "Trust is at issue here," commented Arthur Harrell of the American Council on Transplantation. "Some people are concerned that doctors will prematurely declare them brain-dead. Obviously, we try to allay that fear." Obviously.)

On the Dutch situation, a historical footnote is of interest. The general humanity—indeed, heroism—of the Dutch during World War II was made famous by the story of Anne Frank. Less well known is the story of the Dutch medical profession. When in 1941 Arthur Seyss-Inquart, the Reich Commissar for the Netherlands, ordered physicians to cooperate by, for instance, concentrating their efforts on rehabilitating people who could be made fit for labor, the doctors of Holland unanimously refused.

Seyss-Inquart then threatened to take away their medical licenses unless they cooperated at least to the extent of giving information about their patients to the Occupation authorities. Unanimously, the doctors of Holland responded by handing in their licenses, taking down their shingles, and seeing their patients secretly. They declared that they would not compromise their medical oath, which pledged them to work, solely and always, for the welfare of their patients. Seyss-Inquart argued and cajoled, and then he made an example of a hundred doctors whom he arrested and sent to the concentration camps. But all to no avail. The medical profession of Holland remained adamant. The doctors quietly took care of the widows and orphans of their condemned colleagues, but they would not give in. And so we are told that during the entire Occupation not one of the heroic doctors of Holland cooperated in the Nazi programs of slave labor, euthanasia, eugenic experimentation, and nontherapeutic sterilization.

But all that was a long time ago, and the Dutch doctors of today have so completely forgotten it that the Committee on Medical Ethics of the European Community, in unanimously rejecting the proposals of the Dutch medical society on euthanasia, has expressed "hope that this strong reaction will induce [our] Dutch colleagues to reconsider their move and return to the happy communion of utmost respect for human life."

If the Dutch are being urged to return from the abyss, in this country the forward stampede gains momentum, it seems, almost day by day. This spring, voters in California may have the opportunity to vote in a referendum being pushed by Americans Against Human Suffering, the political arm of the Hemlock Society, which has been around for some years and claims 26,000 members in 26 chapters nationwide. The Hemlock Society's motto is "Good Life, Good Death," and the referendum is promoted under the banner of the "right to die." "We need a public debate on acts of euthanasia, and California has the best track record in the nation for taking unprecedented action," says Derek Humphry, founder of the Hemlock Society. (Mr. Humphry has written a much-acclaimed book on how he provided lethal drugs for his first wife to commit suicide when she had bone cancer.) It is confidently predicted that, even if this referendum fails in California, it will "raise the consciousness" of the nation and open the way to other initiatives.

The referendum, which would legalize active euthanasia and "assisted suicide," is strongly opposed by the California Medical Association. "The public should realize that what we are talking about here is killing people," says Catherine Hanson, the association's legal counsel. "It is absolutely contrary to the entire medical ethic." Proponents of the referendum counter that, in the light of recent developments, such a statement of absolutes is obsolete. They may well be right.

Certainly there has been in the last several years a rash of books, articles, and television programs promoting the "right to die" and, although it is usually not put this way, the permission, even the obligation, to kill. In such advocacy, the linkage is commonly made among abortion, fetal experimentation and exploitation, infanticide, and suicide. The basic argument advanced is the need for rational and scientific control over the untidiness of the human condition.

Among the prominent writers in this campaign are Jeffrey Lyon, Earl Shelp, Peter Singer, Helga Kuhse, and Robert Weir. Singer, for example, has famously argued that your average pig has more consciousness and therefore more right to protection than fetuses or human beings suffering from severe disabilities. (Other animal-rights advocates have exhibited some ambivalence toward this line of argument, knowing that it is human beings, not pigs, that they need to persuade of the rightness of their cause.) In the eugenics literature dealing with issues such as infanticide and suicide, champions of progress typically inveigh against the baneful influence of Christianity in perpetuating irrational "taboos." This would seem to neglect both the proscriptions against homicide in the Jewish tradition and the wondrous flexibility demonstrated by many Christians in accommodating what are thought to be the imperatives of the modern world.

The current eugenics literature is admirably candid about the radicality of what is being proposed. Shelp, for instance, declares that "it is proper to treat unequals unequally," and warns against "a tyranny of the dependent in which the production of the able persons is consumed by the almost limitless needs of dependent beings." Lyon recognizes that many severely handicapped people succeed in living happy, productive, and even inspirational lives. But such people are aberrations ("dynamic, overachieving supercripples") and should not be permitted to distract our attention from the need for a rational public policy that must, perforce, deal with the generality.

In his very useful study, *The Nazi Doctors*, Robert Jay Lifton details the progress of the "medicalization of killing" under the Third Reich. The concept of *lebensunwertes Leben* ("life not worthy of life") was used to cut a wide swath, including the unfit newborn, the mentally ill, the gravely handicapped, the useless aged, and, of course, several races that fell into the category of the "subhuman." It must be acknowledged that, except for tracts issued from the fever swamps surrounding the eugenics project, few people today include a racial factor in calculations of who does and who does not have sufficient "quality of life" to continue living. Here the inhibition against racial discrimination seems to be one "taboo" still firmly in place.

It must be further acknowledged that in the literature there is considerably more moral agonizing about ending the lives of people who have previously been recognized as rational and productive citizens. But in the cases of unfit newborns and human life that is "incapable of full social participation," the decision to terminate is relatively uncomplicated. A rational quality-of-life measurement makes it clear that their lives are not a good for them. Thus the Nobel Prize-winners Francis Crick and James Watson, co-discoverers of the structure of DNA, think that newborn infants should be subjected to rigorous examination and should be permitted to live only if they are found fit. Many who find the proposal repugnant are sure that there is a convincing argument against it, but it does not come readily to mind.

Critics contend, however, that the question of whether life is a good *for* the person gets things backward. The argument of the critics is that life is a good *of* the person, and that depriving the innocent of such life is tantamount to homicide. In current debates, that argument is widely dismissed as "vitalism," which presumably depends upon a metaphysical belief regarding the status of life rather than a rational judgment regarding the quality of the life actually being lived. Admittedly and inevitably, in all cases somebody is making a decision. Such decisions become especially tricky in the instances of involuntary euthanasia or "assisted suicide."

Despite the avid promotion of death with dignity, living wills, and related ideas, the vast majority of people do not, for whatever reason, clearly indicate in advance the circumstances in which they wish to be killed. This results in numerous instances, especially with respect to the biologically tenacious aged, of "subhuman life" being a heavy burden upon family and the medical staff. At this point the assistance offered in assisted suicide must be generously defined, including the decision to make a decision for people who cannot decide. Ethics committees around the country have helpfully developed quality-of-life indexes by which it is possible to make a "best-interest" judgment, also called a "reasonable-person" judgment. That is to say, others decide to terminate a patient's life on the basis of what it is assumed the patient would decide were he a reasonable person acting in his own best interest.

This form of "substituted judgment" has led to concepts such as surrogate suicide or substitute suicide, although of course it is always the other person who dies. Perhaps not surprisingly, when the questions are posed in these ways it is usually decided on behalf of the other party that he or she would decide to stop being a burden to the people who are actually making the decision.

In current practice and discussion, there is not yet a consensus in favor of active euthanasia by administering a lethal dosage or otherwise actually killing the person. A consensus is rapidly forming, however, on withholding food and hydration in order to "facilitate the dying process." This consensus requires the erasure of two distinctions of long standing in medical ethics and practice. The first distinction is between "ordinary" and "extraordinary" means of treatment. It is now widely, although by no means unanimously, agreed that, in the case of certain classes of patients, all treatment is extraordinary, and therefore not required and perhaps not ethically permitted.

The second distinction is between medical treatment and providing food and water. It used to be thought that providing food and water, also intravenously, is a matter of ordinary obligation. The argument is now on the ascendancy, however, that providing food and water constitutes medical treatment. And, again, in specified cases any medical treatment falls into the category of "extraordinary means" which are neither required nor to be countenanced. With the withdrawing of food and water the decision has been made to intervene actively with the clear and sole intent of hastening death. That is to say, the decision has been made for euthanasia or mercy killing. The only question now is how death is to be effected. Starvation is a very clumsy means. The person may live for days, there is often frightful physical disfigurement, and there is the unknown factor of prolonged pain. The attractiveness of starvation to the morally queasy is that it is the "least direct" means of hastening death.

But once we have grown more comfortable with the euthanasia decision that has been made, it seems almost certain that medical practice will adopt means that are more efficient and less aesthetically disturbing. Starvation must thus be seen as a provisional technique to be employed only until medical practice and public opinion are prepared for more rational measures.

It should not be thought that these developments have to do only with the comatose, the "biologically tenacious" drug-sated aged, or others in imminent danger of dying. Those are of course the cases highlighted by euthanasia enthusiasts, for such cases lend themselves to emotionally powerful statements about needlessly prolonging "meaningless" human life, and about the burden that such life is to others. Traditional medical ethics has long allowed the removal of means of sustenance from those near death if the means are counterproductive or ineffectual. In other words, if the feeding instrument is causing other severe disabilities, or the body is not able to assimilate the food, or the person is within hours of dying no matter what is done, intravenous feeding should be discontinued. But what is now being proposed and what is now being done goes much further, including direct intervention to terminate broad categories of people suffering from quality-of-life deficiencies.

The new approach received intense national attention a few years ago in the Baby Doe case in Indiana. There a court allowed parents to starve to death their handicapped baby, even though dozens of couples volunteered to adopt the child. Since then there have been well-publicized cases of adults injured in accidents or suffering from crippling diseases who have been starved to death, although they gave no indication that they wished to die and, at least according to some observers, indicated a will to live. Many questions, of course, have been raised about such cases, most of which are addressed by a recent report from the Hastings Center,

Guidelines on the Termination of Life-Sustaining Treatment and the Care of the Dying.

The panel that issued this report in September 1987 proposes very broad categories of people for whom medical treatment, including the supply of food and water, might be terminated. One category, for instance, is "the patient who has an illness or disabling condition that is severe and irreversible." That would seem to offer distinct possibilities for reducing the population of nursing homes, mental institutions, and a good many hospital wards, thus dramatically relieving pressure on scarce medical resources. The panel focuses on people in such categories who "lack decision-making capacity" with respect to whether they wish to live. In these cases a substituted judgment is required and the "reasonable-person" standard should be applied. The standard is put this way: "Would a reasonable person in the patient's circumstances probably prefer the termination of treatment because the patient's life is largely devoid of opportunities to achieve satisfaction, or full of pain or suffering with no corresponding benefits?" The panel wants it understood that it is being cautious and is sensitive to possible "abuses" of the approach it recommends. Substituted judgments should be carefully reviewed by several parties, including doctors and ethics committees. After listing the several categories of people who are candidates for termination, the report states, "*The above list in no way suggests that treatment should be forgone just because a person falls into one of these categories;* nor does it mean that treatment may not be terminated for other patients." The latter statement, one notes, sharply qualifies and may in some instances nullify the former, despite the former's being italicized. (Treatment, keep in mind, includes supplying food and water.)

Much depends on what is meant by the person's "capacity" to make a decision about whether he wishes to die. "These guidelines define decision-making capacity as: (a) the ability to comprehend information relevant to the decision; (b) the ability to deliberate about the choices in accordance with personal values and goals; and (c) the ability to communicate (verbally or nonverbally) with caregivers."

Any experienced medical "caregiver" will recognize that this constitutes a pretty tall order for many patients. For example, deliberating about choices in accordance with personal values

and goals is difficult for many people under the best of circumstances. Yet the panel urges "respect for the patient as a self-
determining individual" and cautions against "wresting control
from the patient with decision-making capacity." Capacity, we are
told, should not be confused with competence, which is a legal
term. "A person can be legally competent and nonetheless lack
the capacity to make a particular treatment decision." Capacity
turns out to be a marvelously elastic measure. "Capacity is not an
all-or-nothing matter; there is a spectrum of abilities, and capacity
can fluctuate over time and in different circumstances." For instance, "Extreme instability of preference may itself be a form of
decision-making incapacity." The patient who yesterday wanted
to die and today just as intensely wants to live clearly does not
have the capacity to understand what is in his best interest.

The Hastings Center guidelines, which emerged from a
project involving twenty experts over two-and-a-half years, have
been widely hailed. The *New York Times* reported that "experts say
no such comprehensive guidelines have been developed before,"
and the study "breaks important ground." A closer look at the
panel, however, indicates that the document, contra its publicity,
may not reflect such an impressive consensus among experts.

Five of the twenty members of the project, including director Daniel Callahan, are from the staff of the Hastings Center. Of
the remainder, there is a strong representation of people interested in medical malpractice law and of others involved in the administration of nursing homes. Without impugning motives, it
might be suggested that such people have a vested interest in
more relaxed rules for the treatment of people who lack "decision-
making capacity."

In addition, two members of the panel who are ethicists issued
substantive written dissents. Leslie Steven Rotenberg of Los Angeles, who has also publicly challenged the Loma Linda proceedings discussed earlier, is quite forthright: "I fear these guidelines, if
widely endorsed, may be used to give a moral 'imprimatur' to un-
dertreating or failing to treat persons with disabilities, unconscious
persons from whom accurate prognoses are not yet obtainable,
elderly patients with severe dementia, and others whose treatment
is not believed (to use the language of the report) 'costworthy.' "

Despite all this, the Hastings Center report is celebrated as
a landmark document by proponents of the eugenics project, and

is now being invoked in public debates, court cases, and state legislatures around the country.

The director of the Hastings Center, Daniel Callahan, is frequently described as the most widely respected authority on medical ethics in America. Be that as it may, he has certainly been at the center of these discussions for almost twenty years and has recently stirred a lively discussion with his book *Setting Limits: Medical Goals in an Aging Society*.[2] Callahan urges us yet once more to brace ourselves for the thinking of the unthinkable. The basic proposal is that there should be an age limit, perhaps eighty-five, beyond which there will be no government funding for life-extending medical care. Because Callahan is a decent and intelligent man, the proposal is almost painfully nuanced and surrounded by myriad qualifications. Indeed, his is a deeply conflicted and often confused argument. Thus, he offers extensive data indicating that America simply cannot afford quality medical care for a rapidly aging population, but, at the same time, he insists that his proposal should not be adopted for purely fiscal reasons. Again, he repeatedly says that his proposal would be "dangerous" and "morally mischievous" without major changes in cultural attitudes toward aging and death, and such changes, he says, may take generations. Yet he persists in making his proposal now.

Some of the changes advocated by Callahan are surely to be welcomed. Drawing on the work of Leon Kass of the University of Chicago, he urges our accepting the idea that there is such a thing as "a natural life span." In this respect Callahan sets himself against the eugenics project with its delusory dream of immortality through technological control. Yet he simultaneously subscribes to a quality-of-life index by which "natural" limits, such as severe disability, are not accepted but taken to be signs of a life not worth living. Callahan is well aware of the Nazi doctrine of *lebensunwertes Leben* and notes that, in the light of the Nazi experience, "there has been a justifiable reluctance to exclude borderline cases from the human community." That reluctance can be overcome, however, if we keep it firmly in mind that the Nazis "spoke all too readily of 'a life not worth living,'" and if we ourselves are very careful when we speak the same way.

2. Callahan, *Setting Limits: Medical Goals in an Aging Society* (New York: Simon & Schuster, 1987).

Callahan clearly wants to distance himself from the proponents of euthanasia, assisted suicide, and other such measures. But he also argues that "artificial" feeding is a medical treatment and should be discontinued in the case of patients suffering from severe quality-of-life deficiency. Lacking any ethical framework other than liberal individualism, Callahan stresses respect for the patient's decision, or, as it turns out, those who decide for the patient when the patient is "incapable." What it comes down to is quite bluntly stated: "At stake is how far and in what ways we are emotionally prepared to go to terminate life for the elderly."

The sentence is typical of the logic of the eugenics project and interesting in several respects. For instance, it is said that we are terminating life "for" other people, rather than terminating the life "of" other people, it being assumed by the "reasonable-person standard" that we are doing them a favor. As important, we are told that what is at stake is what we are "emotionally prepared" to do. For many people, that is a slight barrier indeed. In this way of thinking, the accent is on freedom, voluntarism, and choice. Nobody is allowed to "impose his values" on others. You are free to decide not to terminate your elderly parent or handicapped child, but you must also agree not to interfere with my decision to "terminate life for" the incapacitated who fall within my decision-making authority. (It is worth noting that the Hastings Center guidelines do include "religious exemptions" for people who have religiously grounded inhibitions about the policies proposed.)

Daniel Callahan is a spirited opponent of the slippery-slope metaphor, insisting that one thing does not necessarily, or even probably, lead to another. But his own emotional preparedness with respect to the treatment of the dependent and incapable has undergone a remarkable development. In the October 1983 issue of the *Hastings Center Report* he wrote forcefully against withdrawing food and water. "Given the increasingly large pool of superannuated, chronically ill, physically marginal elderly, it could well become the nontreatment of choice." He added, "Because we have now become sufficiently habituated to the idea of turning off a respirator, we are psychologically prepared to go one step further." In 1983 Callahan was convinced that "the feeding of the hungry, whether because they are poor or because they are physically unable to feed themselves, is the most fundamen-

tal of all human relationships. It is the perfect symbol of the fact
that human life is inescapably social and communal. We cannot
live at all unless others are prepared to give us food and water
when we need them. . . . It is a most dangerous business to
tamper with, or adulterate, so enduring and central a moral emotion." Four years later Callahan invites us, not to tamper with or
adulterate, but to discard that moral emotion. It is, after all, but
an emotion. One may perhaps be forgiven for thinking that Callahan dramatically illustrates the slippery slope that he so
vigorously denies.

To be sure, there is nothing wrong with changing one's
mind, and people like Daniel Callahan may simply say that they
have thought things through more carefully. As he himself suggests, however, this is not a matter of thinking one's way through
but of feeling one's way through. We need no longer think about
the unthinkable when, in time, it has become emotionally
tolerable, even banal. A useful term in this connection is *primicide*,
the first murder. When it is first suggested that we do a murderous
deed, we may respond, "But that would be murder!" After we
have done it once, or maybe twice, that response loses something
of its force of conviction. As a barrier to evil, novelty is a one-time
thing; it cannot be reinstated. In the 1930s a hit man for Murder
Inc. was on trial. The prosecutor asked him how he felt when committing a murder. He in turn asked the prosecutor how he felt
when he tried his first case in court, to which the prosecutor allowed that he was nervous, but he got used to it. "It's the same
with murder," observed the hit man: "you get used to it."

Champions of the eugenics project are deeply and under-
standably offended when it is said that they are advocating murder. For some reason they do not take offense when the statement
is amended to say that they are advocating what used to be called
murder.

The attempt to deny risk and suffering, the use and elimination
of the unfit—these were all elements of the old eugenics. But what
earlier eugenists could only dream about can now be done; and,
if it can be done, it likely will be done. In the technological possibility of creating "a new man in a new society," we have a vision
that makes the similar ambition of political totalitarians seem
modest by comparison.

Of course there are serious people worrying about that ominous prospect. But it seems that soaring hubris, joined to technical capacity, has broken the bonds of moral restraint. That the bonds are broken is evident enough in the very efforts designed to impose limits.

Thus not long ago textbooks in ethics used to set forth the moral principle that each person counts for one, and none counts for more or less than one. A standard illustration of the principle was the hypothetical case of a hospital with five patients, four of them persons of world-class accomplishment (a statesman, musician, mathematician, and philosopher), the fifth a mental deficient without means or kin. The fifth does, however, have the healthy organs which, if transplanted, could save the lives of the other four. The point was that it could never be right to kill the one in order to save the four, for people are always to be treated as ends and never as means. It was a venerable principle in the history of Western thought. Today the principle is becoming the hypothesis, and the illustration no longer illustrates anything but a "morally agonizing dilemma" to be gravely faced in consultation with surgeons, social workers, ministers, and ethicists.

Or consider, once again, Britain's Warnock Committee. Its chairman, Dame Mary Warnock, flatly states, "There is no such thing as a moral expert." This may suggest that, as a teacher of moral philosophy at Cambridge, Dame Mary is taking her salary under false pretenses, but that is a question for her and her conscience. More immediate to our concern is the assumption that on issues of life and of death, of birth and of the family, "everyone has a right to judge for himself." This is the perfect formula of what Alasdair MacIntyre calls the ethics of "modern emotivism." Step by step, the committee states that, since A is allowed, there is no rational reason for disallowing B. It is, as Daniel Callahan might say, a question of emotional preparedness. Of course the committee knows that some matters of life and death must be regulated by law, but law is a weak reed in the absence of moral reasoning. As Dame Mary writes, "We were bound to have recourse to moral sentiment, to try, that is, to sort out what our feelings were, and to justify them." Most of us, it might be noted, are very good at justifying our feelings.

Studies such as that of the Warnock Committee are not done in a social vacuum. The people involved recognize that they are morally accountable to society, and, we are told, "Society feels, albeit obscurely, that its members, especially the most helpless, such as children and the very old, must be protected against possible exploitation by enthusiastic scientists: and embryos are brought into the category of those deserving protection, just as animals are. This is a matter of public, and widely shared, *sentiment*" (emphasis in original). But the obscure feelings of society are marvelously malleable. So the committee states, "The question must ultimately be . . . in what sort of society can we live with our conscience clear?" That, take note, is the *ultimate* question.

Dame Mary wants it known that she is not unaware of the dangers in this line of thinking. There is, she says, "an increasing sense of urgency" that social controls "should be brought up to date, so that society may be protected from its real and very proper fear of a rudderless voyage into unknown and threatening seas."

And so, according to the Warnock Committee, we have embarked upon this parlous voyage guided by public opinion, technological innovation, and obscure moral feelings, headed toward a society in which we can live with our conscience clear. (It is worth noting that eight of the sixteen members of the committee issued dissents of varying substance. Even so, the Warnock report is hailed as a landmark by the champions of the return of eugenics.)

Of a very different order is last year's document from the Vatican, *An Instruction on Respect for Human Life in Its Origin and on the Dignity of Procreation*. Insisting on the unity of the relational and procreative in human sexuality, the document condemns the new eugenics in no uncertain terms. Procedures such as those countenanced by the Warnock Committee, says the Vatican, are not acceptable. "These interventions are not to be rejected on the grounds that they are artificial" but because they violate the dignity of the human person.

Charles Krauthammer, among others, has treated the Vatican instruction with respect, acknowledging that it is "intellectually more satisfying" than committee products such as Warnock. But he thinks the Vatican statement is also "far less useful." He cites the injunction of the Talmud, "Make ye a fence to the law." A fence prohibits actions that, although not in themselves wrong, open the way to the doing of wrong. The problem with the Vatican statement, says Krauthammer, is that it is "a fence too far." The Vatican, he writes, "sees what hell lies at the bottom of the slippery slope, and rather than erect bulwarks, detours, and sandbags, it declares the entire mountain off-limits." For Krauthammer, "There is no way off the slope." "Better," he asserts, "to find a reasonable way to live on it."

At best, it seems, we can slow the slide to what Krauthammer calls the "hellish center" at the bottom. Reports such as that of the Warnock Committee recommend detailed ethical examination of every inch of our downward slide, and they would even put some provisional obstacles in the way, but their very logic precludes the erection of any fences at all, whether near or far. More than that, they invite the conclusion that there is no hell that the fit and the flexible could not learn to live in with a "clear conscience."

When it comes to the elimination of the unfit, Robert Destro, law professor and member of the United States Civil Rights Commission, believes there might be some safety in the legal tradition and in existing laws. "The prejudice against the disabled and those with mental disabilities," he writes in the *Journal of Contemporary Health Law and Policy*, "is a strong one, with a long and sordid history." In recent years, civil-rights law in particular has been significantly extended to include the handicapped. If courts are now to countenance discrimination against the mentally and physically handicapped by permitting guardians to starve their wards, says Destro, "they should do so directly rather than mask their decisions in high-sounding arguments claiming to rely on 'privacy' and 'self-determination.'" In cases where the ward is incompetent, Destro goes on, the only privacy and self-determination being served are those of the guardian, not those of the ward. On a collision course with the entire history of achievements in civil rights, "the law is in the process of adopting a functional definition of the value of the human person, but it is doing so by indirection." Destro concludes, "Though it may take some time, I do believe that we will live to regret leaving to lawyers, doctors, judges, legislators, and ethicists the important task of deciding who among the disabled shall live, and who shall die. We have been down that road before."

Writing in 1963, Mark H. Haller, a historian of the American eugenics movement, noted that since the war against the Nazis there were signs of "a renewed interest in eugenic problems, although the word eugenics has seldom been used." He cited the noted eugenist Frederick Osborn, who urged the movement to be patient, waiting for scientific knowledge, technology, and social attitudes to prepare the way for the radical changes required. Otherwise, said Osborn, the movement would make the mistakes it did in the past and would once again "turn public opinion back against eugenics."

Twenty-five years later it seems the time is right. Perhaps the law, or maybe the remembrance of horrors past, will yet fend off the return of eugenics in its fullness. Perhaps popular moral judgment, drawn from older traditions of moral truth, will, through the democratic process, begin to erect fences. Perhaps our cultural leaders will rediscover modes of moral reason that appeal to a good beyond emotion. And perhaps not.

And so, quite suddenly it seems, we are facing questions for which we have no ready answers. The questions *are* being answered, however. Most of us, probably because we want to live with a clear conscience, prefer not to think about the answers that are being given. Later, we can say that we did not know.

Part I: The April Conference

To the Athenian Station: Some Reflections on Technical Change and Moral Judgment

Hadley Arkes

In my recent book, I recalled a conversation with the late André Hellegers, who had been a notable fetologist at Johns Hopkins before becoming the director of the Kennedy Center of Bioethics. Both of us had been debating, in print, with a decorated biologist, who had insisted that we cannot know when a human life begins. That question had apparently stirred the deepest puzzlement in Justice Blackmun as well. But André did not think that the problem merited this kind of bafflement among the urbane. It was not arbitrary to speak about death, or the end of life, and André found it no more arbitrary or mystical to speak about the beginning of life. That beginning he placed, without a tremor of doubt, at conception: an entirely new being was created, with a genetic makeup that made the offspring quite distinguishable from either parent. And yet André argued that he was not prepared to regard the zygote, at each of its stages, with the reverence that attaches to human life. He suggested that the zygote may not claim that reverence until it reaches the "chimera" stage, the stage at which it is incapable of splitting and forming identical twins. At that stage it is also incapable of recombining, or merging, with another zygote to form, again, a singular being. Until this stage is reached, the question, André says, is whether we can be sure that the new offspring has yet become "irreversibly an individual, since it still may be recombined with others into one new, final being."[1]

André put it to me in this way: If he and I could touch shoulders and then merge to form a new being—if we could not regain our former separateness, and if we could not be merged any further to create an even newer being—if we had reached, that is, a point of finality in our genesis, then would we not have some doubt about our integrity as separate beings before we had merged? The prospect led me to imagine a case of the following kind. We have two men—we may call one "John" and the other "Martin"—and we may suppose that, when their shoulders touch, they merge to form a unique being of incomparable character whom we may then call "John Martin." And as far as we know, this being has no duplicate anywhere in the civilized world. Let us imagine, further, that John was on his way into town to meet Martin and touch shoulders. But what if John were killed in the street by an assailant before he could meet Martin and consummate his destiny? Would we be inclined to exonerate the assailant because Martin had been incomplete or "uncombined"? Would we be inclined to believe that John had been less than a fully human being with a claim to be protected from unjustified assaults?

What inspired this exercise, of course, was the question posed by intrauterine devices. Those devices work by preventing the zygote from reaching the uterine wall, where the zygote may be sustained on the course of its growth, and where it may make its presence known to its mother. If the zygote en route to the uterine wall is the same new being, with the same claim to our respect, as the zygote implanted, then the operations of the IUD would raise the same moral objections that may be raised about abortions. Even people in the pro-life camp tend to split on this question, and the problem may be so refined as to be incomprehensible to most onlookers. But the question here, in principle, may not warrant all this strain; and it may not accord in its complexity with the odyssey of the zygote through its stages. Here we

---

1. See Hellegers, "Fetal Development," *Theological Studies*, Mar. 1970, pp. 4-5; see also Arkes, *First Things: An Inquiry into the First Principles of Morals and Justice* (Princeton: Princeton University Press, 1986), pp. 408-11.

may be faced simply with a variant of the problem expressed in the line "a funny thing happened to me on the way to the uterine wall." Once we are clear on the nature of the being who is the object of protection, his moral claims are not affected by the fact that he is in transit to a place where he would find it easier to make himself known and to call for protection. In fact, as I suggested in my book, if we created rights and entitlements with the same comic inventiveness that has been shown by the Supreme Court in recent years, we might argue that the person who employs devices or chemicals for the sake of preventing the human zygote from reaching the uterine wall has interfered with the zygote's "right to travel."

There seems to be a tendency in some quarters to settle this question backwards. It seems to be assumed that IUDs and the so-called morning-after pills must be legitimate because they are so common, so readily available. If the protection of the human zygote would make it necessary to rule out these popular devices, that prospect, on the face of it, seems to establish that there is something immanently implausible about the notion of protecting humans from the point of conception. Such assumptions would have to dissolve if the question were addressed in a demanding way, with first matters first. If we were persuaded, for example, that a human being is present from the moment of conception, if we understood that its standing as a human being does not depend on its size or its strength or its verbal ability, then it should make no difference in principle whether the offspring is killed by a bus or a surgeon, by a pill or a coil. The essential moral question will remain: Is this life being taken with or without justification?

That question in principle cannot be affected by the nature of the instruments that are used in taking the lives of newly conceived offspring—or in saving those lives. Whether it is right or wrong to destroy human embryos cannot hinge, then, on the question of whether there is anything we can do, practicably, to protect these embryos and prevent people from destroying them. This is another question that seems to inspire a curious puzzlement in the circles of the educated. Among those people who discuss these matters, one can hear vagrant comments about technology rendering moral choices impracticable, as though the scale of moral judgment was somehow proportioned to the state of our technical arts in any field. Of course, we may be encountering here the kinds of minds that are inclined to believe that, since we can amplify and hear the heartbeat of fetuses far earlier today than we could thirty years ago, the fetuses must be maturing into humans much earlier these days. As professor Daniel Robinson has warned, we fall here into the fallacy of reducing the definition of human beings to the current state of the art in amplifier science.

But anyone who has reflected seriously on the matter would quickly come to see that the nature of the technical instruments would not have a decisive bearing on any question of moral consequence. Let us suppose, for example, that—with the wonders of electronics, computers, and lasers—we were suddenly presented with a new device called Laser Voodoo. With this instrument it would be possible to fix the identity of our victim with precise coordinates and cause him to perish without leaving a trail that would lead back to the electronic hit-man. Things being what they are, we suspect that the first models to appear on the market would be a bit pricey, but several thousand sets would sell at once to those committed buffs who are ready to buy any novel electronic device. After a while, with increased production, the unit costs would come down, and we might be able to buy these instruments at local discount houses. With advances in silicon chips, a device no larger than a keyboard and a small television screen might do the work that formerly required a whole roomful of witches from Haiti. And yet I take it that we would not doubt for a moment that the unjustified homicides carried out with these devices would still be murders.

A melancholy recognition may set in: the instruments are so small, so readily available, both from home and abroad, that it may be impossible to suppress them entirely. We may despair of the possibility of registering our moral condemnation in a law, and yet we soon realize that we have little choice: we know that moral duty does not always accord with self-interest, that many people will discover a compelling interest in removing from the scene people who are getting in their way, and that, for many, the weight of self-interest will overcome any moral inhibitions they may feel about the taking of life. To counter the force of self-interest, it may be salutary here, as in other cases, to add the weight of the law in support of moral restraint. We know, of course, that the law will not be enough. The use of these devices will not be entirely suppressed, and a black market will spring up to supply what the law will not permit in the legitimate marketplace. Some people may worry aloud that we are teaching disrespect for law by seeking to impose on the populace a measure that cannot be fully enforced. These individuals may even be inclined to think that certain people are simply driven to homicide by their natural reflexes, and if they are denied this legitimate device in the marketplace, they may be impelled to seek out back-alley voodoo parlors in shady, unsanitary places, filled with menacing characters. And so we might even find a party springing up to legalize the Laser Voodoo, but seeking to make sure that the operations take place under supervised conditions in government centers with equipment owned by the government. The permission to use these devices may be rationed out to those people bearing notes from physicians claiming that they will suffer acute distress unless they are allowed to dispose of one or two relatives or colleagues who are blocking their careers or making their lives insufferable.

In recent years we have had evidence of widespread evasion of some of our laws on civil rights. The circumvention and the outright flouting of the law seem to be most frequent and pervasive with the laws that bar discrimination on the basis of race in the sale and rental of housing. And yet no one has suggested that the pattern of disobedience and the widespread disrespect for the law have in any way altered the moral premises on which those laws have been founded. Nothing in the record of lawbreaking has inspired legislators to consider anew whether it might not be wrong to discriminate on the basis of race in the sale of housing. If anything, the checkered pattern of compliance has inspired demands for more stringent enforcement. But there are so many transactions in this field that the law will never be perfectly enforced, and we understand that. Still, we have also understood that the law cannot seek perfection in achieving its ends. As Aquinas taught us, the practical purpose of the law cannot be to obliterate all evils but to compress the practice of wrongs. The ambit of lawlessness may be reduced at least to a more tolerable level when wrongdoers appear to be the odd, aberrant characters in a community that does not make vice or injustice a way of life.

But beyond that, the law may have an interest in teaching

certain moral lessons or commending certain standards of judgment to its citizens, and those standards of judgment may be applied by citizens in their private lives, even when they are beyond
the eyes of the law. In *Brown v. Board of Education,* the Supreme
Court articulated a wrong of segregation that seemed to be confined to the experience of schools in a republic.[2] And yet the
country was quickly made sensible to a principle on segregation
that ran beyond the matter of public schools and beyond any
principle the Court had managed to articulate. What was soon appreciated in the country was a principle that barred in a sweeping way—barred in almost all settings—the creation of disabilities or exclusions based on race. Even without the prodding of
the law, people began then to apply that doctrine more broadly,
to housing and colleges and social clubs. They began to see the
rule in all cases as a rule of justice that properly claimed to govern
their private as well as their public lives.

I would bring these lessons back to the problem at hand: at
this moment there seems to be no practicable way of protecting
those new beings who may be done in on their way to the uterine
wall. But that is hardly a reason for suggesting that the issues
raised here are so unfathomable that they elude our moral judgment. By getting past this superstition, we may encourage the exercise of that judgment even in very private settings which the
law at this moment makes no attempt to reach. And in our current politics, these matters will be placed for a long while beyond
the interest of the law. We have no reason to expect that zealous
sheriffs and prosecutors will be seeking warrants to inspect
wombs throughout the land, searching for fertilized ova to
protect. Those who have been concerned with the destruction of
human lives through abortion would be quite satisfied to prevent
the one-and-a-half million abortions that are performed every
year in this country on fetuses that are known to be present. The
opponents of abortion hold back here in prudence: they know that

2. Even as recently as 1983, in the celebrated Bob Jones case, the
Court continued to describe the public policy emanating from *Brown v.
Board of Education* by confining the principle of the case to the problem of
segregation *in schools.* Within the space of two pages the Court referred
five times to the settled public policy that bars "racial discrimination in
education." See *Bob Jones University v. United States,* 76 L Ed 2d 157 (1983),
174-75.

the law as it currently stands makes it legitimate to carry out abortions at every stage of pregnancy for any reason whatsoever. It hardly makes sense for them to expend their capital in opposing intrauterine devices on the grounds that IUDs may destroy embryos, because the law regards the destruction of embryos and fetuses as a thoroughly legitimate undertaking in all of its phases.

An argument over intrauterine devices would not become comprehensible in our public discourse unless the law on abortion were founded on different premises. If the Supreme Court made it legitimate again for legislatures to protect children in the womb, then the public discussion could proceed on different premises: then it might become more intelligible to the public to raise questions about our duty to avoid harm to new beings who may be outside our sight but not outside the range of our concerns or the reach of our moral judgment.

My own hunch is that even then the pro-life movement would be likely to hold back in prudence. Sufficient unto the day is the evil thereof. If the law manages to teach once again that unborn children are human beings, that their lives cannot be taken without a compelling justification, then we may leave the public free to deliberate further from those premises in reaching the more refined questions that arise under the issue of abortion. With the law making the right points and raising the right questions, the public may in time be able to recognize that the children conceived in cases of rape and incest are quite as innocent as the children who are destroyed in other abortions. But even if the public does not manage to surmount its layers of settled prejudice on this particular question, the numbers of abortions involved in rape and incest are but a small fraction of the massive volume of abortions that are performed in this country. The opponents of abortion may be content to live uneasily with the obduracy of public sentiment on such questions. That may seem quite an acceptable price if the same public has come to reject as radically insufficient the grounds on which most elective abortions in this country have been justified. For the pro-life movement may count it as a decided gain overall to have it settled in the public mind that a child does not lose its claim to live because it happens to be retarded, inconvenient, unwanted, or a representative of the wrong gender.

But again, might technical change force the issue? Might

we be overtaken by events that make it harder to settle in with a compromise? The question of the IUD, for example, may be drawn into the public arena as the opponents of abortion are forced to deal with the so-called abortion pill. That pill has been developed and tested in Europe, but so far it has been held back from this country. It has been held back not because of any moral reservations but because of the burgeoning litigation in this country over product liability. The makers of this new pill know that A. H. Robins has been put into bankruptcy as a result of the staggering awards for damages that arose from the Dalkon shield. The climate of litigation has made this a perilous time for the producers of pharmaceutical novelties whose enduring effects on the body have not been fully gauged. As I understand it, the new pill delivers its "effect" in the earliest stages of conception and implantation. Thus the questions it would raise for the law would not be dramatically different from the questions raised by morning-after pills and IUDs. But we have also seen the advent of prostaglandin suppositories, which may be administered by the patient herself in the later stages of pregnancy. Of course, these suppositories have their peculiar "contraindications": the user may face the awkward prospect of delivering the baby alive and at an inconvenient moment—at the theater or the supermarket. If our modern laboratories could immediately produce a pill more powerful and discriminating, the problem of technology and moral judgment could be brought to its finest point of concentration for us. If abortions no longer required the presence of surgeons, paramedics, or clinics; if there was no longer a need for counselors or referrals, for officious prodders and gatekeepers of all kinds; if the result could be accomplished simply with the ingestion of a pill, easily obtained, hardly more menacing in its appearance than an aspirin or a cold capsule—if this were the situation, would it be practicable any longer to legislate against abortion? Would there be any serious prospect for barring access to a pill designed for any woman who wished to have an abortion? And that prospect, already slim, would be reduced even further if a large part of the medical profession clung to sentiments that were out of temper with the law.

But even when we state the problem in this way, in its sharpest form, we may instantly recognize that it is not exactly novel.

In his First Inaugural Address, Lincoln had reflected on the problem of enforcing both the fugitive-slave law and the laws restricting the trade in slaves "in a community where the moral sense of the people imperfectly supports the law itself":

> The great body of the people abide by the dry legal obligation in both cases, and a few break over in each. This, I think, cannot be perfectly cured; and it would be worse in both cases *after* the separation of the sections, than before. The foreign slave trade, now imperfectly suppressed, would be ultimately revived without restriction, in one section; while fugitive slaves, now only partially surrendered, would not be surrendered at all, by the other.[3]

Lincoln was willing to live with a law imperfectly supported as the best means of preserving a restrictive policy on slavery. The immediate object of his policy was not to eradicate slavery in every state in which it was planted but to establish again, as the premise in our laws, that slavery was in principle wrong, that it should not be extended or encouraged, and that it should be placed in the course of ultimate extinction. When those moral premises were planted again in the law, when they took hold in the public sentiment, it would become possible to legislate again, in one discrete step after another, the restriction and dismantling of slavery.

In the same way, the prospects for restricting abortion pills could be enlarged or contracted, depending on the climate of opinion that supports the project of enforcement, and that climate of enforcement would depend in turn on the framework of the laws. Under the current laws, the supports for a policy of restriction would be quite infirm. But, of course, no such policy of regulation may be practicable unless the Supreme Court begins to alter its reigning doctrines on abortion, and that kind of change may well be coming. The path of change would make a considerable difference in shaping the understanding of the public. We would not expect the Supreme Court to turn about overnight and sweep away all legal permissions for abortion, as it once swept away, in a stroke, almost all legal restraints on abortion. Rather,

3. From *The Collected Works of Abraham Lincoln,* ed. Roy P. Basler (New Brunswick, N.J.: Rutgers University Press, 1953), vol. 4, p. 269.

the Court would probably establish that it is legitimate to legislate again on the question of abortion, and that question would be brought back into the public arena. There it would become the object of deliberation in legislatures and public assemblies. And if the laws began to protect again the lives of unborn children, it could only mean that the public had come to believe anew that some abortions may justly be restricted, that the protection of children in the womb may be a legitimate interest of the law. If the sentiment of the public came to settle in this cast, the regulation of abortion pills would take place in a notably different climate of opinion. In that setting, doctors might not be quite so ready to set themselves against the moral temper of the community or mold themselves as adversaries of the moral understanding embodied in the law.

But as we turned to the matter of abortion pills, we would soon be reminded that there is, after all, a framework of law already in place for the regulation of pills or pharmaceuticals. That body of law and regulation already reflects the moral understanding that surrounds the use of pharmacological agents: the druggist does not dispense candy that may be taken at the pleasure of the consumer. The law recognizes the power and danger in these drugs. They may be prescribed only by people of certified competence, who prescribe them for medical need. As Leon Kass has noted, the notion of medical need alerts us to the fact that not everything demanded by a patient represents a medical need which the patient has a right to have met.[4] A man may wish to have a limb amputated in order to honor a bizarre bet. Kass offers the example of a woman who wishes to have a breast removed because it interferes with her golf swing. In the absence of any plausible need or justified end, the wishes of the client do not entail an obligation on the part of a physician to carry out a surgical procedure or administer a drug.

These understandings, as I say, are already contained in a regime of regulation for drugs. Any new abortive drug would presumptively come under this framework of regulation. The burden of argument would fall, then, on anyone who would exempt abortion pills from the same framework of moral questioning,

4. See Kass, *Toward a More Natural Science* (New York: Free Press, 1985), pp. 157-86.

from the same discipline of justification that affects every other drug. Would this pill produce effects any fainter or any more negligible than the effects of the kinds of drugs that are currently subject to regulation? Could a pill be regarded as an innocuous drug if it could destroy a fetus or have lasting adverse effects on the woman who ingests it? Even if we regard abortions as legitimate, it may be a strain to argue that this kind of drug should escape the kind of regulation that is imposed on drugs that are far less powerful and portentous.

But once the abortion pill is again drawn into the framework of regulation, the old questions, the old objections would once more be in place. If there were grounds on which to justify restraints on abortion, there would be grounds on which to justify restrictions on the pill, even if it were easily manufactured and scandalously easy to distribute. The people who sought out abortions might not have to pass by the signposts of abortionists and their clinics—the signs that might stir caution and cause some people to hold back. But the experience would no longer be without strain: there might be the need for clandestine meetings in unprofessional and even shady places, and for transactions that had about them the taste of the unlawful. Again, the laws might be evaded at many points, but those who evaded them would carry the burden of knowing that they put themselves at risk, that they were engaged in an enterprise that was held in contempt by the community. And if that sense of the matter enveloped the trade in pills, it might be, in the end, quite telling.

The late Leo Strauss, a professor of mine, thought that the modern project in science was expressed in that line of Francis Bacon's: that the mission of science was to make itself useful "for the relief of man's estate."[5] In that line Strauss found an expression of the deepest premises of modernity. The interest in self-preservation would be detached from a concern for the moral terms of principle on which our lives would be preserved. The interest in easing life, prolonging it, "relieving" distress and misery, could be pursued as an entirely separate project, detached from the old question of what were the best, or rightful, ends of human life.

5. Leon Kass has made this one of the pervading themes in his thoughtful book entitled *Toward a More Natural Science*.

For the monuments of this modern project, we need merely to look about us. As it happens, the most novel has been the most characteristic: the state of mind cultivated in the modern project has nowhere been reflected more dramatically than in the new technology employing the tissue of fetuses. Nowhere do we find portrayed more sharply that technical ingenuity practiced at the highest pitch of competence and combined with the most serene indifference to questions of moral judgment. For the practitioners in this new field, the moral question is sufficiently satisfied by the prospect of saving or extending lives as a result of these new findings.

The prospects unfolding here may be seen in this way: People suffering from Parkinson's disease may have their tremors eased by the transplant of adrenal glands, which produce dopamine. When the procedure is perfected, there is reason to believe that the preferred tissue will be drawn from fetal brain cells producing dopamine. In West Germany, there have already been successful transplants of fetal organs into children and adults. According to one report in the *New England Journal of Medicine*, three kidneys were transplanted from two fetuses into two children, ages four and nine, and one adult, age twenty-five. In recent years we have seen a growth in the transfer of organs among adults, but it has been estimated by one observer that "a transplant industry based upon fetal-tissue technology could dwarf the present organ-transplant industry." It has been estimated that the transplant of fetal cells might relieve the distress of one million people with Parkinson's disease, 2.5 million people with Alzheimer's, 400,000 victims of stroke, several million diabetics, and several hundred thousand people with spinal-cord injuries.[6]

With these kinds of benefits, the 1.5 million fetuses aborted every year in this country might be viewed in a new way—not as bodies to be discarded but as a field, a vast supply, of medically usable tissue. Right now the state of Massachusetts makes it un-

6. I am relying on Emanuel Thorne's report on the use of human tissue in biotechnical research for the Congressional Office of Technology Assessment. See *The Ownership of Human Tissues and Cells* (Washington: Office of Technology Assessment, 1987), and Thorne's piece in the *Wall Street Journal*, 19 Aug. 1987, p. 16.

lawful for people to sell their fetuses, presumably for the purpose of medical experiments. But if fetal tissue becomes very valuable in the aggregate, that may provide an irresistible incentive to demand a price for a product of one's own body that is so powerfully craved by other people. There has already been a case involving a physician who developed, from the spleen of a patient, a product that may be used in the treatment of cancer. The patient has recently prevailed in his lawsuit: he has succeeded in claiming a share of the royalties that have been received from the marketing of his spleen.

And consider the effects of supply-side economics in this field: the supply may beget a further demand, and we may be on the threshold of a new stage in which women may have the incentive to conceive for the purpose of aborting and supplying the fetal tissue. Even without the presence of a mercantile interest, we have already had a case in which this willingness has been seriously tendered. A woman whose father has been afflicted with renal disease sought to be inseminated with his sperm. Her plan was to have the fetus aborted in the third trimester and have its kidneys transplanted into her father. Evidently, in her case her interest in relieving her father's condition was sufficiently powerful to override a whole ensemble of moral concerns—not only about incest but about the framework of moral relations in which children are conceived and brought into the world. And of course there was the paramount question of whether children are to be ordered up as so many commodities, to be used for their spare parts and then discarded. In this case, permission was refused; the procedure was not carried out. But which state of mind can we really count on to be more dominant in the times ahead? The state of mind that was willing to sweep past a host of serious moral questions for the sake of securing a material good, perhaps saving one's father? Or the state of mind that finally came into play in denying the request?

We are no strangers to the prospect of people willing to sell their blood or their sperm or even to rent out their bodies for the use of others. We might summon a certain obduracy rooted in the past and insist, even in the face of the most powerful incentives of the market, that people may not treat their fetuses as commodities to be sold. Of course, we know that the restraints of the law are bound once again to be imperfect. If some women are willing to be impregnated by the husbands of other women, and if some are willing to be impregnated with the intention of aborting, the law may not be able to cope with all the artifice that is used to disguise the payments. But the law may be able to affect these transactions with a considerable hazard: it may refuse to enforce contracts that provide, in effect, for the purchase and delivery of human beings.[7] As we have already seen, some women change their minds and refuse to deliver the baby they had promised. Still others may decide to keep the baby they had promised to abort. Would we have the courts imposing a compulsory abortion for the sake of upholding a contract? The courts will be compelled to recede from these arrangements, and the entrepreneurs who arrange these affairs may have no way of recovering their down payments. That state of affairs may be discouraging even to the most promising new businesses.

And yet this whole complex of legal restraints and inhibitions is built upon the premise that there is a moral "problem" engaged here. But that sense of a problem may be readily dissolved if one refuses to credit the notion that the bodies of human beings are being sold in this commerce in fetuses and their organs. If the fetus were seen simply as a part of the body, like a growth or a tumor, it might seem no more portentous to sell a fetus and its parts than it appears these days to sell other products of the body such as sperm or blood. Some people tuned to a finer sensibility might nevertheless oppose the traffic in fetal parts for the same reasons that would lead them to oppose the callous dismantling of animals for the purposes of medical research. Still, the laws that forbid this sale of fetuses and their parts would require a firm moral conviction for their support, and that conviction would likely be rooted in the understanding that it is human beings, after all, who are being bred, harvested, and sold. Without that conviction, the regulations of this commerce might be seen as just so many needless, vexing restraints on business. That view of things would be bound to encourage the common reflex to get

7. I offered this argument in relation to "surrogate" mothering at the time of the Baby M case, and this is the argument that seems to have been accepted, finally, by the Supreme Court in New Jersey. See my piece, oddly titled "Judge Sorkow May Have Overstepped His Bounds," *Wall Street Journal*, 9 Apr. 1987, editorial page.

round those laws that vex us to no evident purpose. And if we could not get round those laws very easily at home, we might at least feel less inhibited about using the imports that trickle in from abroad, from countries that might be far more relaxed in regulating this cottage industry in fetal tissue.

Leon Kass has performed the valuable office of pointing out to us the "natural limits" that even science must encounter in extending human life. Certain diseases continue to be absorbing matters of interest to us because they are, as Kass remarks, causes of "unhealth," not because they will cause death. In fact, as Kass has observed, "the complete eradication of heart disease, cancer, and stroke—currently the major mortal diseases—would, according to some calculations, extend the average life expectancy at birth only by approximately six or seven years, and at age sixty-five by no more than one and a half to two years."[8]

Beyond that, Kass has suggested a simple but rather more jolting point: the dramatic gains, the startling improvement in our lives brought about by modern medicine cannot be a *moral* improvement. The "betterment" of our lives may be recorded in several dimensions—the pleasures we live to experience, the children we may see, the symphonies we may write, the applause we may earn—but these goods do not necessarily add up to a moral "betterment." Of course, we may be sustained in our lives to do moral things—to do things that might have been done unjustly in our absence. Certainly our resources for moral understanding have been enlarged by the fact that Plato lived to write his dialogues and Aquinas lived to write his commentaries. But the point is that none of our technical advances adds anything to our moral competencies or our capacity for moral judgment. The melancholy fact, which seems widely to be missed, is that this outcome is immanent in the nature of things. It has to do with the character of moral principles and the capacity that allows us to understand and apply them.

For reasons I have had occasion to explain at length elsewhere,[9] moral principles must be categorical in their logic, and categorical propositions can never be formed about contingent

8. Kass, *Toward a More Natural Science*, pp. 162-63.
9. See *First Things*, chaps. 4 and 5.

things. It could simply never be said, in a sweeping categorical way, that it will always be wrong to use photocopy machines or to sign documents. In contrast, it will always be wrong to hold people blameworthy or responsible for acts they were powerless to affect. Anyone who violates this maxim will be doing something wrong, quite apart from whether he is committing that wrong in a corporation or a family, or whether he is acting in a public or a private setting. We could show that the same thing could be said of anyone who draws adverse inferences about people on the basis of race, and who is moved then to establish disabilities or benefits for people on account of their race. If we understand the principle here, we might notice the engagement of that principle when black people are excluded from schools, from restaurants, from tennis courts, and perhaps even barred from using public photocopy machines. And we would understand that we are not faced here with a series of novel cases: there would be no need to consider whether there is "a right to play tennis" or "a right to use a photocopy machine." The array of cases merely supplies a rich variety of *instances* in which the *same principle* is being manifested.

The parade of cases may offer us a parade of shifting, contingent circumstances, but the moral principle we bring to these cases has nothing to do, at its root, with schools or restaurants, with tennis courts or photocopy machines. If space would allow, we could also show that the principle engaged here enjoys this additional property of categorical propositions: the nature of the wrong involved in these cases is utterly unaffected by the question of whether the wrong nevertheless yields some material benefits or proves oddly productive of happy consequences. For example, what if it were shown in certain cases that the segregation of children on the basis of race actually worked to improve the performance of black children and to raise their incomes in the long run? Nothing in this chain of happy results would affect the wrongness in principle of segregating children on the basis of race. Or at least that is what we would mean to say if we said that racial segregation was *morally*, or *in principle*, wrong. We would understand ourselves to be saying that the wrongness of segregation was not merely *contingent*: that its rightness or wrongness would not depend, in any case, on whether the segregation happened to make black people richer or poorer.

I once recalled, in this vein, the case of Cecil Partee, the prominent black politician and ward committeeman in Chicago. According to Partee's account, he had graduated at the top of his class at Tennessee State University in 1938. As a native of Arkansas, he applied to the law school of the University of Arkansas, but Arkansas had not provided a separate law school for blacks. Rather than admitting Partee to the school restricted to whites, the state offered to pay Partee's tuition at another school, even a private school outside the state. Partee happened to be accepted by the law schools at the University of Chicago and Northwestern University, both notably superior to the law school in Arkansas. Partee finally chose Northwestern, and, as he later commented, "I laughed all the way to Chicago." Partee did not suffer material injury as a result of being excluded on the basis of race from the state law school. For him, in fact, the system of racial exclusion produced a benefit. But Partee was nevertheless *wronged:* he was treated according to the maxims of an unjust principle.[10]

But when we recognize these attributes of moral principles, we understand why they cannot be reduced to things that are merely empirical or contingent. They cannot be reduced to statements about automobiles or luggage or jobs in a pants factory. At the same time, there is nothing vague or nebulous about these principles. There is nothing inscrutable, for example, in the notion of holding people blameworthy or responsible for acts they were powerless to affect. Because the proposition is abstract, it can cover a multitude of concrete cases, but this abstract proposition can also be quite precise in its meaning and in its moral significance. We remind ourselves that moral principles may be enduring, they may be applied to many concrete cases precisely because they are not reducible to statements about empirical things (about automobilies or photocopy machines). As Kant sought to explain, our moral notions are not part of the world of "sensation": the notion of "equity" does not make any impression on our retinas, we cannot see or smell "obligations," and no one has seen "justice" falling out of trees.

I recall these things for the sake of pointing up, again, the distance that separates the properties of moral understanding

10. I recount the story of Partee, along with the attendant explanations, in *First Things*, pp. 95-96 and *passim*.

from the kinds of changes that may be wrought in modern medicine. Even if we managed to produce a species of humans with keener senses—with unimpaired hearing sensitive to a wider range of frequencies, with clearer vision and perhaps even X-ray vision—none of these changes would enlarge our capacity for moral understanding. Through chemistry and surgery, we can tame the furies of an incorrigible, aggressive character and render him notably more pacific. But these new dispositions cannot be counted on to produce reflexes that are always "just." When confronted with a Hitler or with the prospect of the wicked armed, it may be necessary even for the most pacific people to summon the will to resist with arms.

For the same reason, it may not be a moral asset for us to breed humans with the reflexes of watchdogs—men and women who can be counted on to show ferocity toward strangers and who will not shrink from the excitements and the dangers of battle. In that project, the eugenists may hold out prospects as irrelevant to our moral needs as the clumsy offerings of the Skinnerians with their operant conditioning: what they may produce for us, at best, is an Inspector Clousseau who can be depended on to use his club on anyone who comes running out of a bank. But of course, Clousseau is as likely to miss the robbers and assault the bank officials who are giving chase to the thugs. Our operant conditioning, our chemical interventions may alter the reflexes of people, their tempers and inclinations, but these regimens of treatment make no pretense to cultivate within their subjects a capacity to discriminate between justified and unjustified aggressiveness. For that kind of discrimination, they would need some practice in reflecting on the kinds of ends that *justify* the use of violence. Only then might they be able to tell the difference between the excited, agitated men who merit their resistance and the excited, agitated men who merit their support. Between a Himmler and a Patton there may have been little difference in combative temper. The two can be distinguished only by the ends to which they committed their aggressiveness or the political regimes in whose service they concentrated their warlike tempers. The understanding of that distinction cannot well up from the reflexes bred in operant conditioning or from any dispositions that may be induced through the use of drugs.

And yet, let us suppose for a moment that, through the miracle of science, through some combination of secretions or genetic alteration, we could produce people who seem to look at the world through the lens of the categorical imperative. They persistently wonder whether they could "universalize" the maxims on which they have been willing to act, even in the prosaic decisions that make up their days. They may not be able to buy peanuts from a vendor at a baseball game without reflecting on whether they could will, as a universal law, that everyone in the world should be willing to buy peanuts. In other words, nothing in this disposition may rescue them from the most common mistakes that arise when people apply the categorical imperative in the most mechanical, formalistic way, with a thorough want of understanding or imagination. The person who seems to have a gene for the categorical imperative may turn out to be the racist who fancies himself a Kantian because he is fully willing to kill *himself* if he has a black ancestor: he may have come to believe that even a perverse policy can satisfy the requirements of a moral principle if he is willing to install it as a universal law.

Even people who have thought themselves tutored in the categorical imperative have fallen into this mistake. It is not the kind of mistake that can be corrected by tender dispositions or by a passionate concern to be consistent. It can be averted only through serious reflection about the grounds on which we finally judge acts to be right or wrong, justified or unjustified—and the kinds of maxims that *deserve* to be installed as universal laws. That is, of course, a distinctly moral understanding, which requires a certain practice in the discipline of moral reasoning. No improvement in neural mechanisms can be counted on to produce this kind of understanding. Nor is it likely to spring from the transplant of cells or from any alteration in chemistry that markedly improves human disposition.

Winston Churchill offered one of his most searching, thoughtful essays in the early 1930s, when he surveyed the changes produced by science and invention, and looked ahead to the prospects over the next half-century. He called the piece "Fifty Years Hence," and he was moved to a judgment that had to be discordant, or out of temper, with the times. All of this "material progress," he said, "in itself so splendid, does not meet any of the real needs of the human race." What were those *real* needs? Churchill filled in the answer in a roundabout way. He recalled reading, a few days earlier, a book that tracked the history of mankind from the beginning of the solar system to its extinction. Fifteen or sixteen races of men had arisen in succession over tens of millions of years. In the end, he noted, "a race of beings was evolved which had mastered nature": "A state was created whose citizens lived as long as they chose, enjoyed pleasures and sympathies incomparably wider than our own, navigated the interplanetary spaces, could recall the panorama of the past and foresee the future."[11] And yet, Churchill was moved to ask, "what was the good of all that to them?":

> What did they know more than we know about the answers to the simple questions which man has asked since the earliest dawn of reason—"Why are we here? What is the purpose of life? Whither are we going?" No material progress, even though it takes shapes we cannot now conceive, or however it may expand the faculties of man, can bring comfort to his soul. It is this fact, more wonderful than any that Science can reveal, which gives the best hope that all will be well.[12]

Years later, in his memoirs of the Second World War, Churchill would remark that the "main guiding lights of modern faith and culture" had emanated from two ancient cities, Athens and Jerusalem. The main guide to the problems of modern life were to be found in the "simple questions" that were raised for the first time in biblical revelation and ancient philosophy. In a commentary on Churchill, Professor Harry Jaffa has picked out that passage about a new race of humans with "pleasures and sympathies incomparably wider than our own" who can navigate interplanetary spaces. Jaffa surmised that, from Churchill's perspective, the most striking discovery that could arise from interplanetary travel would be the discovery of human life on another planet. Yet, as Jaffa aptly remarked, the discovery of humans elsewhere in the universe would not be merely the discovery of

11. Churchill, "Fifty Years Hence," in *Amid These Storms* (New York: Scribner's, 1932), p. 280.
12. Churchill, "Fifty Years Hence," p. 280.

beings capable of interplanetary travel but the discovery of beings who are capable of asking, again, those familiar "simple questions." But in that case, Jaffa asks, "Why should we search the universe for other beings able to ask these questions, when we can ask them ourselves? What greater mystery is there, or can there be, besides the one within each of our souls? Where can we look with better hope, for a light wherewith to lighten this mystery, than that which came forth, in olden times, from Athens and Jerusalem?"[13]

Churchill seemed to suggest that it is the natural limits of life—the inescapable prospect of death—that compels us to consider the purpose of our lives. That prospect, we might say, concentrates our minds; it forces us to ask ourselves the question, and when we ask it in an insistent way, it may add purpose to our lives. Our colleague Leon Kass has argued in the same vein in his book entitled *Toward a More Natural Science*. In his chapter on the virtues of finitude, he has observed, "the taste for immortality, for the imperishable and eternal, is not a taste that the conquest of aging would satisfy: we would still be incomplete; we would still lack wisdom." We would lack wisdom because we would be diverted from the proper questions and the proper ends. We would be "diverting our aim," as he says, in pursuing bodily immortality; we would be distracting ourselves from the question of what it means to live well.[14] Kass may offer a benign service, then, as he sketches in very precisely the limits of what science can promise in extending our lives. Beyond the sadness of death and its separations, there may be grounds for a new satisfaction in discovering again the simple questions—the questions about the purpose and the justification of our lives—that science cannot dissolve, and invention cannot alter.

13. Harry Jaffa, "Can There Be Another Winston Churchill?" in *Statesmanship: Essays in Honor of Sir Winston S. Churchill*, ed. Harry Jaffa (Durham, N.C.: Carolina Academic Press, 1982), p. 30.
14. See Kass, *Toward a More Natural Science*, p. 314.

Politics and Morality: The Deadlock of Left and Right

Christopher Lasch

Recent controversies about abortion, school prayers, capital punishment, anti-evolution laws, and other so-called social issues have polarized public opinion and sharpened the conflict between Left and Right. Against the right-wing demand for legislative enforcement of morality, the Left insists on the freedom to choose. The Right believes that our society has lost its moral bearings; it blames this development largely on the pernicious doctrines of moral relativism allegedly sponsored by the Left. The Left accuses the right of exaggerating the danger of moral nihilism, dismisses this kind of talk as an expression of "nostalgia," and insists that what looks like moral confusion is better understood as pluralism and diversity: a creative response to unsettling social changes. The real danger, according to the Left, lies in demands for the "imposition" of a rigid, monolithic morality that would force everybody to conform to arbitrary standards that cannot withstand critical examination and therefore have to be upheld by force. The Right worries about chaos, the Left about conformity. Where the Right sees an "abdication of moral authority," the Left sees a "failure of nerve"; those slogans, embodying diametrically opposed interpretations of what is wrong with American society, sum up their quarrel and suggest the futility of trying to resolve it by some process of compromise.

Faced with this deadlock, some people on the Left have begun to advocate a combination of economic radicalism and cultural conservatism. "A few years back," writes Bernard Avishai of MIT, "Senator Paul Tsongas argued that the Democratic party, to move with the times, must become more 'conservative' in economic matters and more 'progressive' in social ones." Tsongas "got it exactly backwards," according to Avishai. Liberals, he thinks, have been too quick to deny the importance and inescapability of moral issues—to equate moral choices with personal preferences and personal or group interests and to adopt a "behaviorist" rhetoric that "seems to make moral debate of any kind obsolete." It is time the Left recognized that "the pursuit of happiness is never entirely a private matter."[1]

The point is well taken, and so is Avishai's point that "discipline, civility, literacy, strength, family, [and] 'faith'" all "deserve a defense from the left."[2] I suspect, however, that a defense of these goods, if it is to amount to more than a half-hearted concession to the growing influence of the New Right, requires a more drastic modification of left-wing ideology than most people on the Left would cheerfully countenance. Avishai puts his finger on the central issue when he observes that "when I say what is good for me, I am inadvertently implying a standard for everybody." But it is hard to see how this view can be reconciled with the principle, so dear to liberals and democratic socialists, that nobody else can tell me what is good for me. John Stuart Mill gave this principle its classic formulation in his essay on liberty, which also gave classic expression to the fear of "conformity": "In the conduct of human beings towards one another, it is necessary that general rules should for the most part be observed, in order that people may know what they have to expect; but in each person's own concerns, his individual spontaneity is entitled to free exercise. . . . All errors which he is likely to commit against advice and warning, are far outweighed by the evil of allowing others to constrain him to what they deem his good." It was on this premise that Mill based his conclusion, the essence of liberalism, that "the only purpose for which power can be right-

1. Avishai, "The Pursuit of Happiness and Other 'Preferences,'" *Dissent* 30 (Autumn 1984): 482-84.
2. Avishai, "The Pursuit of Happiness," pp. 482-84.

fully exercised over any member of a civilized community, against his will, is to prevent harm to others."[3]

A cultural conservatism, so called, that leaves this conclusion unopposed is unlikely to elicit much enthusiasm among cultural conservatives. Nor will conservatives find much sustenance in Avishai's rather anemic list of the "common values" he thinks are "necessary for all of us," including "strong schools, concise language, families that hold together, ... books and public airwaves that do not corrupt our children," and so on. No one can object to such a list, but the question is whether the state should prohibit divorce, say, in an effort to make sure that families "hold together," or whether it should censor material believed to "corrupt our children." Avishai gives a hint of his position on the second of these issues when he hastens to reassure the readers of *Dissent* (in which his article appears) that "we" should not "start joining in calls for federal control of the media or anything like that, though we could, for example, demand more federal money for public television and radio."[4]

Those who advocate a union of cultural conservatism and economic radicalism, a superficially plausible position, necessarily accept the prevailing terms of debate; this is the reason they find themselves drawn to solutions that satisfy neither Left nor Right and leave the present stalemate intact. They ask themselves, in effect, how we can counter moral chaos without the risk of conformity. When the issue is posed in these terms, it is tempting for those who find themselves at odds with both Right and Left to seek some middle ground between these extremes. Consider the Bergers' book entitled *The War over the Family*, which announces its intention in the subtitle, *Capturing the Middle Ground*. The Right, as the Bergers see it, wants to dismantle the welfare state but at the same time to give the state vast powers over morality. The Left wants to strengthen the welfare state but opposes attempts to enforce a moral consensus. It condemns the family as both pathological and obsolete. The Bergers argue, on the contrary, that the family remains the basic social institution, providing a source of emotional stability in a changing world and equipping the young with the

3. Mill, *On Liberty* (Buffalo: Prometheus Books, 1985 [London, 1859]), pp. 16, 17, 87.
4. Avishai, "The Pursuit of Happiness," pp. 483-84.

qualities of self-reliance and selfdiscipline on which a democratic society has to rest. Their unapologetic defense of "bourgeois values" obviously puts them at odds with the Left, but their insistence on the need for a welfare state, together with their opposition to moral rearmament, defines their disagreement with the Right.[5]

The Bergers recognize the class bias that informs current disparagement of the family. They understand that the "war over the family" is a class war. They see it, however, as a war between the "new class" and the middle class (defined so broadly that it now includes most workers as well), whereas it is better understood as the latest phase in a long-standing conflict between middle-class individualism and working-class tribalism. According to the Bergers, the antifamilial ideology adopted by the new class represents a betrayal of "bourgeois values," but in many respects it merely makes the underlying implications of those values explicit. It aims to complete the struggle against tribalism by liberating women from the family and integrating them into the market, by protecting the rights of children, by discrediting "pronatalism" and upholding the ideal of planned parenthood, and by removing every obstacle to the individual's pursuit of his own private advantage.

The logic of individualism cannot be easily arrested, as the Bergers seek to arrest it, at a point in its development that defines the family as a private living arrangement but still insists on the need for strong families. The family is strong precisely to the extent to which it is not a purely private arrangement. Once the family comes to be defined as a haven in a heartless world, it becomes increasingly difficult to resist the conclusion that sexual activity is purely a transaction between consenting adults and that it is "*not* a function of the state," as the Bergers themselves declare, "at least in a democracy, to regulate the arrangements by which 'consenting adults' arrange their private lives." The Bergers im-

5. Brigitte Berger and Peter L. Berger, *The War over the Family: Capturing the Middle Ground* (Garden City: Anchor Press, 1983), p. 172: "The family, and specifically the bourgeois family, is the necessary social context for the emergence of the autonomous individuals who are the empirical foundation of political democracy." See also p. 205: "It is the *bourgeois* family . . . whose primacy we want to be recognized [their italics]."

mediately qualify this statement by adding that they "would not want to be confused with those liberals who have taken the position that all morality is a private matter and thus of no concern to government." Every society needs a "basic moral consensus," they believe, but since they refuse to countenance any "efforts to impose, by law, an essentially provincial morality on the entire society," it is not clear how they expect such a consensus to declare itself. A common allegiance to "reliability, honesty, industriousness, respect and concern for others, willingness to take on responsibility," and other virtues pertaining to "interpersonal relations in the private sphere" hardly amounts to much of a "moral consensus," nor can it nourish a "sense of community."[6] This is the liberal morality of live-and-let-live, and it is incapable of commanding loyalty to anything beyond the individual's immediate interests. The ability to command loyalty—and we should take this phrase quite literally—is the mark of a community, together with a willingness to enforce it even when loyalty seems unreasonable. The morality of tolerance, on the other hand, frees individuals to do as they please as long as they allow others the same freedom. It implies that all preferences and "lifestyles" are equally valid but equally devoid of the qualities that would evoke lifelong devotion.

Suppose the most important danger in our society is neither chaos nor conformity but apathy—a certain "want of faith," as Emerson used to put it, an attenuation of the capacity for loyalty and devotion, an absence of demanding standards of conduct that would elicit ardor and enthusiasm.[7] The characteristic vices of a commercial society remind us that the real question is character, not social order or individual freedom or the attempt to balance the two. In his well-known essay entitled "Post-Modernist Bourgeois Liberalism"—still another attempt to break out of the present impasse of political debate—Richard Rorty argues, in effect, that liberal intellectuals have weakened the idea of loyalty by

6. Berger and Berger, *The War over the Family*, pp. 76, 147, 184, 206.
7. "The disease with which the human mind now labors is want of faith. . . . We renounce all high aims. We believe that the defects of so many perverse and so many frivolous people who make up society, are organic, and society is a hospital of incurables" ("New England Reformers" in *The Complete Essays and Other Writings of Ralph Waldo Emerson*, ed. Brooks Atkinson [New York: Modern Library, 1940], p. 458).

identifying it with the "common interest of humanity" and by appealing to "transcultural and ahistorical" standards of justice. "Loyalties and convictions" cannot be effectively defended by an appeal to "general principles," according to Rorty. They are rooted in the "beliefs and desires and emotions" of particular groups, "with which we identify for purposes of moral or political deliberation." It was easier for American intellectuals to understand that loyalty to their own community was "morality enough," Rorty thinks, in the days when they "still thought of their country [as] a shining historical example." The Vietnam war, however, caused intellectuals "to marginalize themselves entirely" and to take the position not merely that the war "betrayed America's hopes and interests and self-image" but that it "was *immoral*, one which we had no *right* to engage in in the first place." The effect of this appeal to universal principles, Rorty points out, was "to separate the intellectuals from the moral consensus of the nation rather than to alter that consensus."[8]

Elsewhere Rorty deplores the "collapse of moral self-confidence" in a culture whose "sense of its own moral worth is founded on its tolerance of diversity." According to Rorty, tolerance has deprived us of "any capacity for moral indignation, any capacity to feel contempt." It has dissolved our very sense of selfhood. Under these conditions, it might be a good idea to remind ourselves that our national beliefs, however parochial, are "worth fighting for," and more generally that tribalism—"exclusivity," Rorty calls it—is a "necessary and proper condition of selfhood."[9]

Rorty's defense of parochialism recalls the defense of endangered cultures issued by Claude Lévi-Strauss. Instead of condemning ethnocentrism—the anthropologist's stock-in-trade—Lévi-Strauss upheld it as the "price to be paid so that the systems of value of each spiritual family or each community are preserved and find within themselves the resources necessary for their renewal." Invited to open UNESCO's International Year to Combat Racism and Racial Discrimination, Lévi-Strauss elected to challenge his audience in its comfortable belief in tolerance,

8. Rorty, "Post-Modernist Bourgeois Liberalism," *Journal of Philosophy* 80 (1983): 584, 586-88.

9. Rorty, "On Ethnocentrism: A Reply to Clifford Geertz," *Michigan Quarterly Review* 25 (1986): 526, 532-34.

universal brotherhood, and the "family of man." He "rebelled,"
he said, "against the abuse of language by which people tend
more and more to confuse racism" with a willingness to value our
own way of life "above all others." This kind of loyalty, Lévi-
Strauss argued, was "not at all invidious." "All true creation implies a certain deafness to the appeal of other values." Against the
modern "delusion" that "equality and fraternity will some day
reign among human beings without compromising their diversity," Lévi-Strauss insisted on the importance of "relative incom-
municability." In order to retain their vitality, he thought, cultures
had to remain "somewhat impermeable toward one another."[10]
It would be hard to find a more explicit repudiation of the conventional point of view among social scientists and among anthropologists in particular, a viewpoint exemplified by Franz
Boas's lifelong crusade for tolerance, understanding, and recognition of the unity of mankind.

But what if exclusiveness and tribalism breed animosity, intolerance, and irreconcilable social conflict? Moral fervor is all
very well, but what if it leads all too quickly to fanaticism? The
great achievement of modern liberalism was to rid the Western
world of religious warfare, and the most convincing defense of
liberalism has always been the claim that a renewed outburst of
religious fanaticism (now attached to secular ideologies like
communism and anti-communism) remains an ever-present
danger. As is well known, liberalism does not take a very
generous view of human nature. It sees mankind as essentially
acquisitive and self-seeking, and it therefore regards its own victory over intolerance and fanaticism as a provisional and highly
precarious achievement.[11] The fear of uniformity is misplaced,

10. These passages from two of Lévi-Strauss's essays—"Race and
Culture" and "The Anthropologist and the Human Condition"—are
quoted in Clifford Geertz, "The Uses of Diversity," *Michigan Quarterly
Review* 25 (1986): 106-8.

11. The only way to understand politics, according to James Tully,
is to "recognize, as Locke did in 1660 [in his early manuscripts, *Two Tracts
on Government*], that modern politics is the continuation of war by other
means and, as he saw in 1667 [in the early drafts of the work published
in 1689 as *A Letter concerning Toleration*], that the only means that keeps
it from converting back to war . . . is toleration" (Introduction to Locke's
Letter concerning Toleration [Indianapolis: Hackett, 1983], p. 16).

according to liberals. The last thing we have to fear, writes Clifford Geertz, is a world "full of people so passionately fond of each other's cultures that they aspire only to celebrate one another." In opposition to Lévi-Strauss, Geertz warns against an "easy surrender to the comforts of merely being ourselves." The growing recognition that thought, instead of serving as a mirror of reality, operates only in concrete communities of discourse, each with its own particular conventions and traditions, does not mean that "human communities are or should be semantic monads, nearly windowless." The discovery that the "limits of my language are the limits of my world" ought to be construed as an invitation to master new languages, not as a reflection on the impossibility of communication across cultural barriers.[12]

In his attempt to meet this line of argument, Rorty's own argument takes an unexpected turn—one that demonstrates, once again, how difficult it is, even for those who challenge the dominant epistemological conventions, to break out of the conventional terms of political debate. The danger that tribalism will lead to intolerance and social conflict can easily be reduced, according to Rorty, by giving up the misguided political theory that "makes assent to ... certain moral ideals a requirement for citizenship." It is the yearning for a "world polity whose citizens share common aspirations and a common culture" that generates conflict, since it leads each culture to try to bring the others around to its own way of thinking. When persuasion fails, competing cultures inevitably turn to force. Their mistake lies in equating social order with moral consensus, democracy with brotherhood. If only they would "disassociate liberty and equality from fraternity" and renounce the "old-timey" ideal of *gemeinschaft*, they would appreciate the modest but valuable achievement of liberal institutions in "allowing individuals and cultures to get along together without intruding on each other's privacy, without meddling with each other's conceptions of the good." The incommensurability of these conceptions may rule out "community" in the "strong approbative sense" of the term, but it does not rule out Rorty's "postmodern bourgeois liberalism"—postmodern because it has rid itself of the delusion that society needs a consensus about values or that such a consensus can ever be achieved.

12. Geertz, "The Uses of Diversity," pp. 110, 113, 122.

Democracy requires procedural justice, according to Rorty, not moral consensus. Liberalism brought an end to religious warfare by the simple expedient of divorcing church and state, and the same solution—the privatization of moral beliefs—is equally applicable to the growing diversity of twentieth-century societies.[13]

Geertz agrees with Rorty that "the world is coming . . . to look more like a Kuwaiti bazaar than like an English gentleman's club."[14] Elaborating on these images in his exchange with Geertz, Rorty urges a very different way of dealing with cultural diversity—not bilingualism but the "construction of a world order whose model is a bazaar surrounded by lots and lots of exclusive private clubs." In the guise of something daringly "postmodern," Rorty resurrects the classic formula for peace in liberal societies: "after a hard day's haggling, retreat to your club," where you will be "comforted by the companionship of your moral equals." The only surprising feature of Rorty's argument is his eagerness to export liberalism—"Rawlsian procedural justice," he calls it—to the rest of the world. A defense of particularism might seem incompatible with the contention that liberalism represents the "best hope of the species"—not just a way of life appropriate for the West but a way of life that can be recommended enthusiastically to others, perhaps even forced on them if necessary. Rorty insists, however, that as long as our recommendation implies no claim of truth or moral superiority we can safely admit to ourselves that liberalism is "worth fighting for." As an ideology designed for export, liberalism may be as "parochial" as any other, but it has the unique advantage, according to Rorty, that it asks very little of its adherents—only an acknowledgment of the pointlessness of public arguments about truth, morality, or religion. Africans and Orientals do not need to love the West or abandon any of their own cherished ideals in order to recognize "procedural justice" as the only workable solution to the problem of diversity:

> One does not have to accept much else from Western culture to find the Western liberal ideal of procedural justice attractive. The advantage of postmodernist liberalism is that it recognizes that in recommending that ideal one is

13. Rorty, "On Ethnocentrism," pp. 533-34.
14. Geertz "The Uses of Diversity," p. 121.

not recommending a philosophical outlook, a conception of human nature or of the meaning of human life, to representatives of other cultures. All we have to do is point out the practical advantages of liberal institutions.

In the context of his argument, Rorty's assertion that our "ideals of procedural justice" are "worth fighting for" seems to carry the disturbing suggestion that fighting is somehow less "intrusive" than debate. The really unforgivable "intrusion," it appears, is to try to change someone's mind.[15]

Rorty's attempt to explain how we can "fight" for our beliefs without "intruding on each other's privacy" or "meddling with each other's conceptions of the good" is unintentionally instructive. It helps us to understand the curious way in which liberalism promotes the ideological conflicts it is supposed to discourage. What makes liberalism such a seductive but also such a pernicious ideology is its refusal to recognize itself as an ideology. It insists on its own neutrality, claiming to represent nothing more than a set of procedures. Its superiority allegedly lies in its indifference to ideology. But of course liberalism, like other ideologies, implies a "philosophical outlook," a "conception of human nature," and a set of assumptions about the "meaning of life." Those assumptions emerge clearly enough in Rorty's description of the public realm as a bazaar, where the absence of moral consensus does not prevent us from "haggling profitably away." What distinguishes liberalism from other ideologies is not the absence of assumptions about human nature and the meaning of life but its refusal to formulate those assumptions explicitly and thus to submit them to the test of public debate. Under the guise of tolerance, liberalism banishes the exploration of moral and religious questions to "private clubs," where the companionship of our "moral equals" serves to strengthen the unearned conviction of superiority, an easy "capacity to feel contempt," in Rorty's phrase.[16] What liberals forget is that when we try to change other people's minds, we run the risk of changing our own. This is the difference, after all, between

15. Rorty, "On Ethnocentrism," pp. 532-34. For Geertz, on the other hand, the position advocated by Lévi-Strauss and Rorty is objectionable above all because it forecloses the "possibility of quite literally, and quite thoroughly, changing our minds" ("The Uses of Diversity," p. 114).
16. Rorty, "On Ethnocentrism," pp. 526, 533.

argument and warfare. "Meddling" with others' "conception of the good" may alter our own conception of the good. Because our age is unduly impressed with the incommensurability of competing conceptions of the good life, we find it difficult to think of conversation (political conversation in particular) except as warfare carried on by other means, and our determination to respect others' opinions leads to the ironic outcome that warfare comes to seem less "meddlesome" than argument itself.

The very activity that unites us with people otherwise different from ourselves—the effort to change their minds—is perceived by our age as the source of fatal divisions. Our determination to respect others' opinions actually withholds the one essential mark of respect: a willingness to argue with them. The refusal to argue not only betrays a lack of respect for others but also undermines self-respect, because it leaves our own opinions untested and therefore shaky and undisciplined. It is not enough to hold strong opinions; the only way to find out whether they are good for anything is to submit them to cross-examination. Recent controversies about the role of "values" in education suggest that both the Left and the Right find it hard to grasp this point. Both sides in these debates tend to confuse advocacy with "indoctrination." The nature of the educational enterprise, one might think, lies precisely in the distinction between the two. Whereas education seeks to elicit belief through persuasion, those who have no confidence in education rely on coercion or manipulation, often disguised as "education." A good teacher knows that conviction achieved through coercion or manipulation is easily shaken as soon as it is exposed to counter-manipulation. Good teaching consists of leading students through all the objections to a given position, the expectation being that any position worthy of loyalty will withstand rigorous questioning.

Good teaching has become increasingly rare, however. Perhaps because American intellectuals no longer regard their country as a "shining historical example," most teachers suspect that the old-fashioned educational program, with its emphasis on the training of character, conceals a "hidden curriculum," as Lawrence Kohlberg once put it.[17] Convinced that "values" mean unthinking

17. Kohlberg, quoted in Edwin J. Delattre and William J. Bennett, "Where the Values Movement Goes Wrong," *Change* 11 (Feb. 1979): 39.

obedience to authority, uncritical patriotism, and unqualified acceptance of the American way of life, they try to avoid "*teaching values* . . . in the sense of promoting or inculcating values," in the words of Robert Hall, and to confine themselves instead to "*teaching about values* in the sense of helping young people to appreciate the fact that certain values are commonly held in the society in which they live." The "dilemma" of moral education, as Hall sees it, is to avoid the extremes of relativism and indoctrination. No one has found the "*via media.*"[18] This formulation sounds plausible until one sees that it dissolves the distinction at the heart of the whole practice of education, the distinction between indoctrination and persuasion. The direction of this kind of argument becomes unmistakable when Hall rebukes Kohlberg for upholding the Rawlsian theory of justice as the highest stage of moral development, reproving Kohlberg not on the grounds that Rawls's theory is open to devastating objections but on the grounds that Kohlberg should not uphold any position at all. In doing so, Kohlberg "approaches indoctrination," according to Hall.

The same premise—that advocacy and indoctrination are indistinguishable—underlies Kohlberg's reply to Hall, one of his last contributions to the debate on moral education. Belatedly aware of the dangers of moral relativism, Kohlberg happily accepts Hall's charge of indoctrination. Having renounced his "negative views of indoctrinative moral education," Kohlberg now argues that "there is a developing 'rational' or 'natural' sense of justice that is universal to all, . . . an awareness of 'natural law' in light of reason." It is therefore permissible for schools to urge children not to lie or steal but not to promote the "religious and metaphysical doctrines" that might give these injunctions some depth and resonance. Religious and metaphysical doctrines, it appears, are unsupported by the "natural" sense of justice. "Schools may teach *about* religious and metaphysical doctrines," says Kohlberg, "but not with the intent of their teaching's being accepted by students as the truth."[19]

Opponents of "secular humanism" rightly object to this

18. Hall, "Moral Education Today," *Humanist* 38 (Nov./Dec. 1978): 12-13.

19. Kohlberg, "Moral Education Reappraised," *Humanist* 38 (Nov./Dec. 1978): 14-15.

kind of talk, but they themselves seldom challenge the assumptions behind it. Convinced that the public schools neglect instruction in basic subjects, leave children morally confused, and fail to uphold high standards of conduct, they might welcome Kohlberg's second thoughts about falsehood and theft and still insist that more drastic forms of indoctrination are required. Children need to believe in God and country, and it is up to the schools to instill belief—if necessary, by preventing exposure to contrary beliefs. The religious Right argues further that parents, after all, ought to have the last word about what their children are required to learn. On the other hand, it is an article of faith among "humanists" that "respect for the parent's rights," in Kohlberg's words, has to give way to "respect for [the] child's autonomy."[20]

"People without convictions cannot be counted on," it has wisely been said.[21] The question is why so few Americans seem to have any convictions that don't crumble at the first exposure to contrary convictions. What should we make of the well-known evidence that American POWs seem unusually susceptible to brainwashing, evidence often cited by both sides in the debate about values in education? Does it mean that our schools are failing to inculcate an unshakeable sense of national pride? Or does it mean, as liberals argue, that schools should teach children how to think instead of trying to teach them what to think? Unfortunately, they are unlikely to learn how to think unless they have something to think about. Skills without content will not do much for anybody. But neither will an education that aims merely to reinforce the parental point of view. If the school serves merely as an extension of parental authority, teachers will lose interest in their work, hobbled by parentally imposed restrictions on freedom of thought. Parents have a right to insist that schools should respect local customs and beliefs. But the attempt to insulate children from knowledge that calls those beliefs into question betrays a want of confidence in their capacity to withstand criticism. To forbid teaching about socialism, say, does not help children to un-

---

20. Quoted in Kathleen M. Gow, *Yes Virginia, There Is Right and Wrong! Values Education Survival Kit* (Toronto: John Wiley, 1980), pp. 46-47.

21. Delattre and Bennett, "Where the Values Movement Goes Wrong," p. 42.

derstand what might be valuable in capitalism, any more than proscription of evolutionary theory makes them better Christians. The attempt to ban subversive books and subjects only strengthens educators in their determination to make the curriculum as bland and inoffensive as possible, to offer a little of everything to everybody, to subordinate content to method, and to give equal time to every point of view.

The trouble with our schools is not that children are learning secular humanism and moral relativism instead of patriotism and Christianity but that they aren't learning much of anything. They are being bored out of their minds by an education the only consistent aim of which is to avoid giving offense. Left-wing pressure for schools that leave everything open to question and right-wing pressure for schools that leave nothing open to question combine to bring about the same sorry result—schools in which nobody thinks about anything at all. From the Left, educators hear dark warnings of a "retreat to authoritarian habits," as Ellen Goodman puts it in a diatribe aimed principally at William Bennett's proposal for a common curriculum.[22] From the Right, they hear proposals to stop moral decay by supplementing the theory of evolution with "creation science." The attempt to accommodate both kinds of criticism leads to a system of education the most outstanding feature of which is apathy. The same thing is true of politics, that larger school in which people have become accustomed to hearing only what politicians think they want to hear. The reluctance to offend anyone and the relegation of the most important issues to the seclusion of "private clubs" have drained all the moral passion out of politics, just as they have drained the passion out of education.[23]

22. Goodman, "A Tale of Three American High Schools," *Rochester Democrat and Chronicle*, 23 Jan. 1988.

23. Compare Rorty's contrary view that "American universities are in better shape than any other institutions of learning in history." Against Allan Bloom's insistence (in Rorty's paraphrase) that "we have to ask not just about institutional procedures but about the substance of what is being taught," Rorty argues that knowledge should be thought of as a "bag of tools." The university should be thought of as a "flea market," and "students should be left free to shop around in as large and noisy a bazaar as possible" ("That Old-Time Philosophy," *New Republic*, 4 Apr. 1988, p. 32).

Our despair of politics originates in the belief that a vigorous civic life is possible only when people share the same ideals, the same cultural, ethnic, and racial background, even the same interests, and that without these unifying influences, politics tends to degenerate into warfare. We define as the prerequisites of politics the very conditions that would make politics unnecessary—cultural homogeneity and universal agreement. In the absence of agreement, we seem to be confronted with two equally unattractive alternatives. On the other hand, we can agree to disagree—in private. In other words, we can agree to keep divisive moral issues out of politics, even though this means treating political questions as technical questions of economic management, fiscal policy, and social engineering that should be left to experts. This solution has a number of undesirable effects, one of which is that it relegates the rest of us to a passive, spectatorial role. If, on the other hand, a fully developed civic life depends on moral consensus, we can try to create such a consensus by means of indoctrination, proscribing subversive books, enforcing school prayers, and beefing up agencies of law enforcement in an attempt to keep people in line. Both programs lead to the same result: the suppression of public debate. Both arise from the failure to hold morality and politics in tension, a failure that leads to a dichotomy: either politics has everything to do with morality or it has nothing to do with morality.

The conflict between Left and Right, insofar as it is a conflict about the relation between morality and politics, is rooted in the old argument between *gesellschaft* and *gemeinschaft*—an argument that remains interminable because a common premise makes the two positions mere reflections of each other. The Right clings to the appealing but historically misleading vision of a tightly knit little world in which everyone agrees on a common definition of the good life; it thereby lays itself open to the charge that any attempt to reconstruct this kind of common purpose in the modern world leads to ideological conformity and eventually to totalitarianism. Liberalism at least acknowledges the diversity of opinions and interests that precludes the creation of a republic of virtue—but liberalism is no more satisfactory than its communitarian mirror if it issues only in the relegation of morality to private life. The privatization of morality deprives us of any common life at all and makes politics a battleground on which conflicts can be resolved only by superior force.

It would seem that we need a conception of politics neither communitarian nor individualistic, a conception best described as fraternal. Fraternity recognizes the boundary between the self and others, but it does not despair of crossing it. It does not take refuge in the "capacity to feel contempt" or seek to fortify the self against dependence on others, to achieve a state of complete self-sufficiency. Nor, on the other hand, does it try to annihilate the self in the hope of achieving universal brotherhood. As Wilson Carey McWilliams says, and as many others have said in different words, "It is possible to love everyone equally only if one loves nothing in particular."[24] The ability to love something in particular, someone in particular, rests on a refusal to see others merely as extensions of ourselves. But this doesn't rule out the hope of a common civic life. On the contrary, it is the only thing that makes a common civic life possible, because it creates the possibility of trust. A love of the particular, in other words, makes us see that the circumstances of our collective insecurity in the world make it necessary to trust those who cannot be subjected to our control, treated as instruments of our will, or brought into perfect agreement with our own views and purposes. A politics of fraternity recognizes that the impossibility of complete agreement makes it all the more essential to expose moral issues to political debate. It conceives of trust not as a favor we bestow on those whose purposes coincide with our own—an attitude that would drain the concept of trust of its meaning—but as a necessity imposed on us by our common weakness.

24. McWilliams, *The Idea of Fraternity in America* (Berkeley: University of California Press, 1973), p. 48.

The New Eugenics and Feminist Quandaries: Philosophical and Political Reflections*

Jean Bethke Elshtain

> *It is especially the female body, with its unique capacity to create human life, which is being expropriated and dissected as raw material for the industrial production of humans. For us women, for nature and for the exploited people of this world, this development is tantamount to a declaration of war. For us women it means a further step towards the end of self-determination over our bodies, our ability to procreate and consequently our ultimate dependence on medical experts. We declare that we do not need or want this technology and that we fight it for what it is: a declaration of war on women and nature.*

> Bonn Conference: Women against Gene Technology and Reproductive Technologies, April 1985

*I dedicate these reflections to my daughter, Sheri, who has triumphed over such tags as "minimal brain dysfunction," "cerebral deficit," "behaviorally disadvantaged," "mentally challenged," and "mildly retarded" by describing herself, simply, as someone who "thinks different." Also, thanks to my graduate assistant, Tobi Elkin, who undertook bibliographical searches, photocopied materials, checked out books, and discussed with me the questions I explore in the pages to follow.

Normative heterosexuality must be replaced by a situation in which the sex of one's lovers is a matter of social indifference, so that the dualist categories of heterosexual, homosexual and bisexual may be abandoned. . . . We must remember that the ultimate transformation of human nature at which socialist feminists aim goes beyond the liberal conception of psychological androgyny to a possible transformation of "physical" human capacities, some of which, until now, have been seen as biologically limited to one sex. This transformation might even include the capacities for insemination, for lactation and for gestation so that, for instance, one woman could inseminate another, so that men and nonparturitive women could lactate and so that fertilized ova could be transplanted into women's or even into men's bodies. These developments may seem farfetched, but in fact they are already on the technological horizon; however, what is needed much more immediately than technological development is a substantial reduction in the social domination of women by men. Only such a reduction can ensure that these or alternative technological possibilities are used to increase women's control over their bodies and thus over their lives, rather than being used as an additional means for women's subjugation. Gayle Rubin writes: "We are not only oppressed as women, we are oppressed by having to be women or men as the case may be."

Alison M. Jagger, *Feminist Politics and Human Nature*

This essay is not an examination of in vitro fertilization, AID (artificial insemination by donor), embryo flushing, surrogate embryo transfer, surrogate motherhood, sex preselection, cloning— the entire panoply of real or potentially realizable techniques for manipulating, redirecting, controlling, and altering human reproduction—so much as it is a consideration of assumptions that underscore or undermine alternative feminist positions toward these developments, which are frightening (according to one view) or potentially hopeful and helpful (according to another view). The two quotations that begin this essay offer some inkling of just how vexing the entire matter of radical intrusion into human biology and our ideas, and ideals, of embodiment, childbirth, human intimacy, and intergenerational ties has become to feminists. The Bonn Conference represents a radical *noninterventionist* stance; the position spelled out by Jagger, a feminist philosopher, is radically *pro-interventionist*, so much so that it foresees technological elimination of male and female. There are, of course, feminists ranged in between these two polar stances.[1]

But the perplexities deepen as one explores various feminist positions, for one discovers that there are often ways in which an *explicit* stance—say, opposition to surrogacy—is undermined by a tacit commitment to a framework imbedded in ontological presumptions that erode one's political refusals. Needless to say, all this needs to be parsed out, and I will attempt to do so in a way that does not bog down in analytic finery but remains on ground accessible to all thoughtful and reasonably well-informed readers.

It is necessary to set the stage for current debates—including the way in which such highly visible controversies as the Baby M case stirred up so much turmoil inside feminist circles—by locating modern feminist positions within a discursive and historic tradition. The dominant forms of contemporary American feminism are heavily indebted to a stance I shall tag *ultra-liberalism*. One need not espouse all the features of ultra-liberalism in order to be held captive by it. It is in the air we breathe; it is, as Wittgenstein might put it, a *picture* that holds us captive, "and we cannot get outside it, for it lay in our language and language seemed to repeat it to us inexorably."[2]

No one among us can shed her cultural skin altogether. But each of us can become more reflective about what makes our social and political order what it is. One can reconceive relations and practices, perhaps to reveal a richness and ambiguity that received formulae conceal. One aim of the social theory I embrace—

1. Anne Donchin, "The Future of Mothering: Reproductive Technology and Feminist Theory," *Hypatia* 2 (Fall 1986): 121-37, divides up feminist positions into non-interventionist, moderate interventionist, and radical interventionist. Our tacks on these issues are somewhat different, my approach being more philosophical and explicitly lodged in moral concerns than Donchin's. But her discussion is helpful, and each of us rounds up many of the usual suspects.

2. For the next few pages I draw upon an earlier essay, "The Liberal Captivity of Feminism: A Critical Appraisal of (Some) Feminist Answers," which appeared in *The Liberal Future in America: Essays in Renewal*, ed. Philip Abbot and Michael B. Levy (Westport, Conn.: Greenwood Press, 1985), pp. 63-84. Since only about two dozen people in the world seem to have discovered this book, I don't feel too bad about recycling several arguments.

an interpretive, desimplifying enterprise—is to work toward a more complete description of what is really going on, of how things are with us. This may be done, or attempted, in a number of ways. One might expose certain practices and doctrines as self-defeating or, alternatively, rescue other practices and ideals as vital to a cherished vision of the human community. No doubt one cannot require of political activists that they be self-critical and reflective about their enterprise from its inception—given the heat of battle, the demand for solidarity, and the urgency to get results. So they proceed: "We hold these truths to be self-evident." Self-evident truths are often bulwarks, common rallying points, and powerful symbolic markers. But they may rapidly become liabilities, locking social participants into usages that point them down false paths to liberation or wisdom. Such seems to me the story of feminism and the doctrine of ultra-liberalism.

The reigning ultra-liberal frame cannot be stretched to cover every form of feminist association and identity. My hunch is that many grass-roots feminists, absorbed in immediate and practical concerns like health care, sanctuary for battered women, school curriculum questions, and so on, have not troubled themselves overmuch about their underlying philosophy. And that philosophy—again a hunch—would probably emerge as a pastiche, a combination of maternal and community imperatives, including care and sharing, plus evoking "rights" and "choice" as absolutes, or nearly so. The new eugenics puts terrific pressure on feminists who are not ultra-liberals in particular ways, just as it compels those in the ultra-liberal camp to push the envelope about as far as it can go, to come out explicitly for total human control over "nature"—but only if women control that control. I will say more about this shortly, but first I want to further unpack the dominant philosophic presumptions of ultra-liberalism and the stance toward our embodiment they tend to yield—after I lodge the usual cowardly caveat.

I do not share the view of some critics who proclaim our liberal inheritance a rotten deal through and through. Liberalism is not of a piece. By distinguishing ultra-liberalism from liberalism as such, I hope to avoid painting a monochromatic picture. Although I believe there are sources for political renewal within liberalism, I also find these options under pressure to succumb to the combined force of those beliefs, practices, and commitments I shall criticize. This is a worry. If I am even partially correct, it means that the dominant liberalism, its priorities and doctrines deployed increasingly as vehicles and rationalizations for newer modes of social control, has become self-defeating. Inside this picture, one is stuck with more of what has helped to sicken us in the first place: the final rationalization and disenchantment of all aspects of social life; a deeper dependency of the self on anti-democratic bureaucracies and social engineering—and now genetic engineering—elites; a more complete stripping away of the last vestiges of personal authority (construed as domination), traditional identities (construed as irrational and backward), and so on. The entanglement of contemporary feminism with *this* liberal project is a feature of the social order that too often goes unmarked, perhaps even unthought.

FEMINISM AND THE ATOMIST TURN IN POLITICAL DISCOURSE: BACKGROUND CONUNDRUMS

What makes ultra-liberalism run? The foundational motor that moves the system is a particular notion of the self. That self helps to make ultra-liberalism what it—complicatedly—is, including its characteristic construal of social reality. There is no single, shared understanding of the self that grounds all forms of liberal theorizing. The transcendental subject of Kant's deontological liberalism, for example, is a being at odds with the prudential calculator of Bentham's utilitarianism. Ultra-liberalism's vision of the self flows from seventeenth-century atomism, a doctrine linked to the names of Hobbes and, less tightly, Locke. Atomism posits the self as given, prior to any social order—ahistorical, unsituated, a bearer of abstract rights, an untrammeled chooser in whose choices lie his (and initially this character was a "he") freedom and autonomy.

One uneliminable feature of atomism, then, "is an affirmation of what we could call the primacy of rights."[3] Although atomism ascribes primacy to rights, it denies the same status to any principle of belonging or obligation. Primacy of rights has

---

3. Charles Taylor, "Atomism," in *Powers, Possessions and Freedom: Essays in Honor of C. B. MacPherson,* ed. Alkis Kontos (Toronto: University of Toronto Press, 1979), p. 39.

been one of the important formative influences on the political consciousness of the West. We remain so deeply immersed in this universe of discourse that most of us most of the time unthinkingly grant individual rights automatic force: in our political debates, rights are trumps. Atomism makes this doctrine of primacy plausible by insisting on the "self-sufficiency of man alone or, if you prefer, of the individual."[4] Closely linked to the primacy of rights is the central importance that atomists attach to freedom understood as "freedom to choose one's own mode of life," to constitute and choose values for oneself.[5] In making freedom of choice an absolute, atomism "exalts choice as a human capacity. It carries with it the demand that we become beings capable of choice, that we rise to the level of self-consciousness and autonomy where we can exercise choice, that we not remain mired through fear, sloth, ignorance, or superstition in some code imposed by tradition, society, or fate which tells us how we should dispose of what belongs to us."[6] Solidified by market images of the sovereign consumer, this atomist self was pitted with great success against the self of older, "unchosen" constraints.

Thus we find gender preselection presented as a choice, a consumer preference, or an "option."[7] Surrogacy is lodged in the language of contractual rights, rights so sacrosanct that they override the right of the surrogate to change her mind and keep the baby after birth—or so 61.8 percent of the respondents in a Gannett News Service–*USA Today* poll claimed in the aftermath of the Baby M trial. Mr. Stern's rights to enforce a contract were favored three to one, indicating that when I write of a society entangled with the atomist project I speak not just of elites but of a wider population.[8]

4. Taylor, "Atomism," p. 41.
5. Taylor, "Atomism," p. 48.
6. Taylor, "Atomism," p. 48.
7. Laurie Bobskill, "Couples in Western Massachusetts Offered 'Gender Preselection,' " *Springfield Union News*, 4 Dec. 1987, p. B-7.
8. Interestingly, the higher the educational level among respondents, the more the Sterns were favored over Mary Beth Whitehead. The same held for income level, of course, since education enjoys a strong positive relation to income. Also, when gender is used as a differentiating category, more women than men favored the Sterns (61.3 percent to 54.8 percent), indicating, I believe, how strongly the "best interests of the

This atomist picture of freedom remains so deeply entrenched that we tend to see the natural condition or end of human beings as one of self-sufficiency. Atomism's vision of the self, its absolutizing of choice, and its celebration of radical autonomy all cast suspicion on ties of reciprocal obligation or mutual interdependence and help to erode the traditional bases of personal authority in family and polity alike. Feminists make their indebtedness to atomist construals powerfully manifest when they proclaim choice an absolute, granting the "right to choose" *prima facie* force. The likely result is that any perceived constraint or chastening of individual choice is suspect and will be assessed from the standpoint of the atomist standard. And there is another outcome. Once choice is made absolute, important and troubling questions that arise as one evaluates the writ over which individual right *and social obligation*, respectively, should run are blanked out of existence. One simply gives everything—or nearly everything—over to the individualist pole in advance. Of course, embracing free choice and primacy of rights as self-evident truths makes sense, for, as the daughters of liberal society, women *have* been deprived of freedom understood in the ways I have just described.

Moving from the plane of abstract discussion to a particular issue, feminist images of *the body*, deepens our awareness of the background of dilemmas for feminism that the new eugenics presents. In feminist debate the female body has been constituted as the locus of a struggle for control. To appreciate the terms of feminist description and evaluation of the body, whether in early radical feminism, liberal feminism, or much of Marxist feminism, one must first look back at liberalism's historic suspicion of the body and its desires. Classical liberalism valorized a public world in which adult (male) persons, stripped of particular passion, shared an identical commitment to prudential reason. The arbitrariness of desire was consigned to the non-public sphere; it lay outside the official writ of the liberal *episteme*.[9] Yet this poten-

---

child" argument has taken hold—and that the character assassination of Mary Beth Whitehead by Judge Harvey Sorkow had some effect.

9. *Episteme* is a concept drawn from Michel Foucault. It refers to "the total set of relations that unite, at a given period, the discursive practices that give rise to epistemological figures, sciences, and formalized systems." See Foucault's *Archaeology of Knowledge*, trans. Alan Sheridan-Smith (New York: Harper & Row–Colophon, 1972), p. 119.

tially chaotic and uncontrollable desire was required to hold liberal rationalism intact, serving as its mirror opposite. Keeping the beasts in their place, milling about inside the private corral, liberals aimed to liberate men from a double subjection: to the rule of the traditional patriarch and to privatized passion.[10]

John Stuart Mill accepted this bifurcation, contrasting Reason with the abyss of Instinct, "the worse rather than the better parts of human nature." The rational man, striving for an "apotheosis of Reason," must reject utterly the "idolatry" of Instinct, an idolatry "infinitely more degrading than any others."[11] Only when relations between men and women are lifted up to the realm of Reason, Mill declares, will they be free from the taint of unscrupulous desire. The body itself must come under the sway of the wider social force of rationalization and its ethos, one that dictates setting to one side, or stripping away, all distinctions and specific identities that situate us and define us as particular beings rather than universal, abstract agents. And it is true that to make good on Mill's version of the liberal promise, men and women must abstract *from* their sexual identities and eschew "arbitrary passions"; only then can they usher in the halcyon world of sexual equality.[12]

Occlusion and denial of the body became a dominant thread in modern feminist discourse. Mill's distaste gave way to Simone de Beauvoir's disdain. Her animus, though extreme, makes discursive sense within the framework of Sartrian existentialism. Beauvoir's excoriation of the female body as "inessential," a prefixed abyss that condemns women to a netherworld of unfreedom (Immanence), bearing the stigma "victim of the species," is by now familiar to social theorists. Beauvoir constituted the body as an alien Other, the enemy of the free project of Transcendence. The woman, given her biological capacity to bear a child, is by definition an alienated being. Menstruation, childbirth, nursing—Beauvoir portrayed them all as aspects of a chamber of bodi-

10. For a complete discussion, see my book entitled *Public Man, Private Woman: Women in Social and Political Thought* (Princeton: Princeton University Press, 1981).

11. Mill, *The Subjection of Women* (Greenwich, Conn.: Fawcett, 1970), p. 18.

12. Mill, *The Subjection of Women*, p. 141.

ly horrors.[13] Sharing a central marker of atomism, Beauvoir split the rational self from its unfree (if one is female) embodiment.

Beauvoir helped to set the terms for a feminist repudiation of embodiment that reached its apogee with Shulamith Firestone's 1972 tract, *The Dialectic of Sex*. Her indebtedness to Beauvoir made explicit, Firestone located the oppression of women *in nature:* we are oppressed *because* we are embodied. Swallowing whole a depiction of nature as unfree and unthinking—lacking sentience and meaning—Firestone embraced atomism, technocratic hubris, genetic engineering, and a rather muddled notion of aesthetics as her feminist utopia.

The question for us now is not whether these are extreme voices but whether, in perhaps extreme form, they point to broader forces at work in, and on, feminism. As the scrambling for position over the new eugenics in its many manifestations shows, our embodiment remains a serious predicament for feminists.

DAMAGE CONTROL: HOW FEMINISTS ARE CONFRONTING THE EUGENICS CHALLENGE

Although feminist discourse from the mid-sixties on was lodged securely in the notion of reproductive freedom, few feminist thinkers paid much attention to newer technologies for controlling human reproduction other than to issue briefs on behalf of a one-hundred-percent safe and effective form of contraception and on behalf of no limits to abortion-on-demand. The voices from within the feminist camp that questioned arguments for abortion couched *exclusively* in the language of absolute freedom of choice, or rights, did not prevail in the debates. These voices now seem prescient, given the runaway developments in reproductive technology and genetic engineering of the past decade. How, then, are feminists responding?

For one group, the *radical pro-interventionists*, the new eugenics presents no problem so long as it can be wrested from male control. Early radical interventionists construed technology

13. See my essay entitled "Existentialism and Repressive Feminism," in *Liberalism and the Modern Polity*, ed. Michael J. Gargas McGrath (New York: Marcel Dekker, 1979), pp. 33-62.

instrumentally—it would help to usher in a wholly new social order, a feminist utopia, so long as feminists struggled towards this end (though their historic teleology dictated that this end was well nigh irresistible). Thus Firestone offered us the cybernetician as the modern savior. Her scenario for salvation went like this: (1) free women from biological tyranny; (2) this freeing will undermine all of society and culture that is erected upon biological tyranny and the family; (3) all systems of oppression—including the economy, the state, religion, and the law erected upon the family—will erode and collapse. The woman who absorbed Firestone's position could work to attain this promised future by *seizing control* of reproduction and *owning* her own body. Ultimately, Firestone continued, test-tube babies would replace biological reproduction as the chief means of reproduction, pregnancy having been declared "barbaric" along with the "fat lady" peered at by strangers, laughed at by children, and deserted by feckless husbands. Full victory would be achieved when every aspect of human life rested in the beneficent hands of a "new elite of engineers, cyberneticians."[14] The child, with no need to be "hung up" by authoritarian parents (parents having pretty much melted away), is free to bargain for the best deal in contracted households.

Interestingly, although some feminist theorists criticized Firestone from the moment her book appeared, it is only with the advent of the new eugenics that her work is being treated with rather widespread skepticism—if it is mentioned at all. Thus Anne Donchin writes,

> Though Firestone's advocacy of technological reproduction aims to serve feminist interests, it rests on conceptual foundations that have much in common with the presuppositions of researchers and policymakers who would . . . support technological intervention for the sake of the monopoly of power it would make possible. Both sorts of interests view technology as "a victory over nature." . . . Both see human biology as a limitation to be overcome—for Firestone, because she takes the relations of procreation to be . . . the source of women's oppression;

14. Firestone, *The Dialectic of Sex* (New York: Bantam Books, 1972), p. 201. See also Elshtain, *Public Man, Private Woman*, pp. 205-28.

for those who would support "a brave new world," because the diffusion of power among women and families threatens their own power hegemony.[15]

At the heart of the radical interventionist position is an insistence that biological "tyranny" must end and the corollary demand that biological sex and social gender can and must be sharply severed—at least until such time as these categories have altogether disappeared. The only possible opposition to the new eugenics that might emerge from the pro-interventionist camp takes the form of warning that evil forces—masculinists, anti-feminists—are controlling the means of control. Thus an ongoing political struggle is required to insure that patriarchalists do not succeed in this effort. But pro-interventionist caveats have been compromised by the fact that they *share* rather than *oppose* all the ontological assumptions of their opponents: that "nature" must be overcome; that where human beings find the will to indulge such acts of transcendence, they must find a way; that only the fearful and the backward will cavil at these inexorable developments—again, with feminist interventionists insisting that if women are in charge, the outcomes will be beneficent.

I find nothing in the feminist pro-interventionist literature that offers up a strong, principled objection to a claim such as the following from Peter Singer and Deane Wells:

> If the creation of new forms seems a godlike power, what more noble goal can humanity have than to aspire to it? Like Prometheus, the mythical Greek hero who defied the gods and stole from them the secret of fire, should we not challenge the gods and make their powers our own? Or to put it in more scientific terms, should we allow ourselves to remain at the mercy of genetic accident and blind evolution when we have before us the prospect of acquiring supremacy over the very forces that have created us?[16]

Seeing in women's links to biology, birth, and nurturance only the vestiges of our animal origins and patriarchal control, anything

15. Donchin, "The Future of Mothering," p. 130.
16. Singer and Wells, *Making Babies: The New Science and Ethics of Conception* (New York: Scribner's, 1985), pp. 157-58. Indeed, Singer and Wells use Firestone to insist that feminists support their position.

that breaks those links is by definition to be applauded. Firestone's celebration of a technological resolution to woman's "control deficit" portrays absolute control over reproduction as the "final freedom"—a position she shares with non-feminist pro-interventionists like Singer and Wells. But they are now on the defensive, at least tacitly, as feminist apprehensions have risen.

RISING CONCERN AND THE GREENING OF FEMINISM

In June 1979 a workshop called "Ethical Issues in Human Reproduction Technology: Analysis by Women" was held at Hampshire College in Amherst, Massachusetts. The workshop was structured around several important issues: contraception, sterilization abuse, prenatal diagnosis, neonatology, sex preselection, and such "manipulative reproductive technologies" as in vitro fertilization, egg fusion, and cell manipulation. The organizers aimed for a "women-centered analysis," defined as "total: physical, mental, emotional"—in contrast to "male-centered" principles of domination, objectification, exploitation, hierarchism, and profit. Foreseeing the final resolution of *all* value conflicts once women's values came to prevail, one of the conference organizers optimistically looked forward to "real solutions" to the many dilemmas of the new technology.

Divided between the bad "them" and the good "us," the conference paper-givers—with few exceptions—presaged more recent developments pitting patriarchalists against women/nature. Tellingly, only one participant raised the question of eugenics—claiming that women shouldn't really be talking about "amniocentesis and similar techniques without putting it in the context of the search for 'bad genes.'" Ruth Hubbard went on to remind participants that the eugenics which had its heyday in the latter part of the nineteenth and the early part of the twentieth century (and which was supported by such birth-control advocates as Margaret Sanger) died because its techniques were imperfect. But our techniques have improved—so must our vigilance, she insisted. One participant challenged feminist use of gender preselection as a way to eliminate male fetuses—following artificial insemination—as "stupendously sexist." But, on the whole, participants celebrated women-centering and the need for "women control" of reproductive technology.[17]

The ten years since that conference have witnessed an efflorescence of feminist anti-technological efforts concentrating on "nuclearism" and war on the one hand and biology on the other. The powerful voices to emerge are more "green" than "red," more separatist than integrationist—that is, they believe women must stand apart from the principles of patriarchal society rather than seek equality on a par with men. The *radical non-interventionists* now dominate the debate, and other feminists are compelled to elaborate their own positions in relation to those of the radicals. Seeking to salvage the principle of women's right-to-choose and wrest it from men's attempts-to-control, the non-interventionists have fashioned a potent rhetoric and conjured up terrifying scenarios—inviting, or catalyzing, moral panic on the part of readers who are, as I am, sympathetic to many of their concerns.

Continuing to "demand the right to choose," the non-interventionists are pondering the nature of the many coercive choices that the new reproductive technology seems to produce. Is amniocentesis really a free choice, or is it a manipulative and coercive procedure with only one *correct* outcome—to abort if the fetus is "defective"? What about the "right" to a child? Is it absolute, or is this, too, yet another imposition of patriarchal society upon women who see themselves as failures if they cannot get pregnant? In addition, the values identified with mothering (even under patriarchal controls) are now being re-assessed as feminists are encouraged to experience maternity in a "non-exploitative way."

The anti-interventionist argument, as articulated by its most radical proponents, shares with the pro-interventionist stance the assumption that all of human social life and all of history is patriarchal. The difference is that the anti-interventionists hold that all modern technology is designed explicitly to deepen and extend patriarchal control and "masculinist" patterns of thought. They are deeply skeptical that *this* technology can be turned to good purposes. Thus nuclear arms cannot be controlled; they must be eliminated. And thus nearly all new forms of reproductive en-

17. See my article entitled "A Feminist Agenda on Reproductive Technology," *Hastings Center Report* 12 (Feb. 1982): 40-43.

gineering cannot be reconfigured to meet feminist and women's needs; they, too, must be deconstructed. Working from the analogy of prostitution, radical non-interventionists insist that just as males moved successfully to control female "sex parts" through various forms of prostitution (including marriage), so they seek a new reality: the reproductive brothel. Writes Andrea Dworkin: "Women can sell reproductive capacities the same way old-time prostitutes sold sexual ones. ... . While sexual prostitutes sell vagina, rectum, and mouth, reproductive prostitutes will sell other body parts: Wombs. Ovaries. Eggs."[18] (Needless to say, this position is condemnatory of all men and contemptuous of many women.)

Gena Corea, perhaps the most visible North American feminist anti-interventionist and founder of FINRRAGE (Feminist International Network of Resistance to Reproductive and Genetic Engineering), insists that the patriarchal state reduces "women to matter." She portrays men as having such total control that they compel the choices "women learn to want to make." In her book, *The Mother Machine,* she draws a scary portrait of present and future horrors as she moves from the farmyard to the bedroom, showing the ways in which methods first developed as part of animal husbandry are making their way—all part of partriarchal plots, in her view—into human lives.[19] It would be too easy to dismiss all the arguments being made by non-interventionists because their underlying philosophic assumptions are so dubious: women = nature = good; men = anti-nature = bad. But it is hard to figure out how to pick and choose.

Here are a few suggestions. Hard-line feminist anti-interventionism can be questioned not so much because of what its proponents *oppose* but because of the reasons proffered for this op-

18. Dworkin, cited in Genoveffa Corea, "How the New Reproductive Technologies Could Be Used to Apply the Brothel Model of Social Control over Women," *Women's Studies International Forum* 8 (1985): 299.

19. Gena Corea, *The Mother Machine* (San Francisco: Harper & Row, 1985), pp. 2, 134, 233. One reason that the new reproductive technology could burst upon the scene with such apparent suddenness is the fact that many decades of intrusive and cruel experimentation on animals went unmarked. Animals' bodies have been violated for decades as part of a quest for scientized production and control. There is a direct line between laboratory experiments on animals, farm factory production, and the new eugenics.

position. The arguments turn on no developed moral position concerning the nature of the human community and moral responsibility, proponents preferring to focus instead on dubious equations of women with nature and hence on women as the only true source of creativity. In addition, by continuing to assert an absolute right-to-choose but *only so long as* the choices are *true* choices, not *false* ones, the anti-interventionists promote a world in which good girls (women-identified women) must fight not only patriarchs but also bad girls (male-identified women). Thus the lesbian who wants to assert her right to "independent motherhood" has every right to artificial insemination. But the woman in a heterosexual relationship who, with her husband, opts for in vitro fertilization is a hapless dupe of patriarchal wiles. This won't do.[20]

One can share the apprehension of anti-interventionists concerning eugenics but deepen and expand their worries as part of an alternative philosophy. I read document after document condemning the new reproductive engineering *because* it meant patriarchal society would seek to eliminate women through genetic manipulation—a rather wild idea. Screening out and eliminating imperfect fetuses also got targeted by many anti-interventionists as a patriarchal perversion. Yet most do not want to interdict the possibility, although the decision to abort a "defective" fetus must be solely that of the mother and made in a noncoercive setting. But this early detection and selective elimination of the imperfect unborn relies on the very technology that the anti-interventionists condemn as patriarchal. To be genuinely com-

20. I oppose the invasive technology of in vitro fertilization because it deepens the atomistic worldview, it enhances the demand for technological solutions as well as the power of those who proffer such solutions, it disjoins human intimacy and reproduction, and it promotes the view that human beings have an absolute right to their *own* child rather than sustaining them in accepting the limits of their own fertility while at the same time encouraging them to develop a loving relationship to children who are not their direct biological offspring. But the women and men who seek, often desperately and repeatedly, to reproduce biologically deserve empathy, not contempt. To call such women prostitutes is cruel. Cf. Christine Crowe, "Women Want It: In-Vitro Fertilization and Women's Motivations for Participation," *Women's Studies International Forum* 8 (1985): 547-52. Crowe questions "choice" and shows the ways in which IVF provides for technological intervention in procreation in ways that mirror power relations between males and females as groups.

pelling, the anti-interventionists would have to extend their opposition to eugenics to include gender preselection by feminists undergoing AID (artificial insemination by donor). Either one does have or one does not have moral permission to eliminate the unborn on the basis of gender. But anti-interventionists will not broaden their opposition in this way because a preferential option for the female fetus is part of the arsenal of weapons to fight patriarchal society.

The radical anti-interventionists are right to insist that "technical progress is not neutral." But to counterpose good women's values to bad male values gets us nowhere. There are women as well as men who support these technologies—some in the name of feminism. One anti-interventionist insists,

> We can no longer subscribe to the technocratic utopia of Bebel and all the other scientific socialists who think that the liberation of women will come with the electrification of the kitchen (Bebel) or with microprocessors or even through technical "liberation" from the biological process of childbirth (Firestone), in short, by the further "development" of "productive forces" plus socialism. . . . The so-called new technology does not bring us and our children any kind of qualitative or quantitative improvement in our lives, it solves none of our basic problems, it will advance even more the exploitation and humiliation of women; therefore we do not need it.

This strikes a very sympathetic chord with many who do not share the full panoply of anti-interventionist assumptions.[21] And warning flags are going up in unexpected places, including the *Village Voice*, which featured a piece on "the selling of in vitro fertilization" in which the author indicates that "tears of gratitude" sprang to her eyes when a Catholic priest on an ethics panel mentioned "conjugal intimacy"—the only person to do so in a week-long discussion of reproduction that was otherwise "desexed, disembodied, dehumanized."[22]

21. Maria Mies, "Why Do We Need All This? A Call against Genetic Engineering and Reproductive Technology," *Women's Studies International Forum* 8 (1985): 559.

22. Andrea Boroff Eagan, "Baby Roulette," *Village Voice*, 25 Aug. 1987, p. 16.

MODERATE INTERVENTIONISTS AND QUEASY FEMINISTS

We have traversed enough territory for the reader to appreciate the python-like grip of the atomistic worldview, of ultra-liberalism, on the thinking of interventionists and anti-interventionists alike. The more consistent (though deadlier) position of the interventionists—whether of the feminists I have discussed or of the many more plentiful technocrats, "pharmacrats," scientists, and profit-makers—is thoroughly saturated with the ontological presumptions of this potent *Weltanschauung*. The radical anti-interventionists feature a farrago of ultra-liberal and pre-liberal, romantic-expressivist presumptions linking women and nature: I have in mind the hard-core greens. But most feminists and most people generally (or so I would guess) belong fitfully in between, hopeful that real help might come to infertile couples but in ways that seem human and humane, concerned to "do something" about human suffering but worried about eliminating human beings who seem, to many eyes, to *be* suffering by definition. (I refer to children born with spina bifida or Down's syndrome or any number of other congenital conditions.) Most folks support contraception and do not want abortion made illegal—but neither are they "pro-abortion."

Take, for example, a report issued by the National Council of Churches of Christ (USA) that offers a "cautious but positive stance toward the emerging technology." Casting human beings as co-creators, the report builds on a particular interpretation of "dominion" from Genesis and Psalm 8, going on to claim that Scripture "exalts the idea that men and women are coming into the full exercise of their given powers of co-creation" and citing genetic engineering as the case in point.[23] Or take the position of Ruth Hubbard, cited earlier, who opposes eugenic and fiscal arguments for abortion but supports particular women who may want to "avoid bearing a child with a disability"—in light of an often unsupportive wider social surround.[24] One finds a shaky combination of *sic et non*.

23. Ronald S. Cole-Turner, "Is Genetic Engineering Co-Creation?" *Theology Today*, Oct. 1987, pp. 344-45.
24. Hubbard, "Prenatal Diagnosis and Eugenic Ideology," *Women's Studies International Forum* 8 (1985): 568.

The Baby M case crystallized this queasiness and prompted further refining of what might be called the moderate position. (Just how moderate is, of course, a matter for debate.) Here was a situation in which biological motherhood and social parenting were severed—as feminists had long been claiming they should be. Here was a situation in which a father insisted he wanted to father—as feminists had long claimed men ought to want to do. Here was a case in which everyone freely agreed to a contract. Not surprisingly, the upsurge in feminist opposition to commercial surrogacy and Judge Sorkow's decision flummoxed several well-meaning sorts, including one David Lipset, who wrote a letter to the *New York Times* in favor of "orthodox Western feminism," which, he claimed, *must* support surrogacy as a way to lay "to rest the old Freudian saw that women were biologically destined to do little other than mother."[25] Why were feminists perversely reaffirming the importance of biological motherhood, he opined—or, rather, pined?

A good question. Betty Friedan claimed that the initial decision denying Mary Beth Whitehead *any* claim—she was no mother of any kind in any way—had "frightening implications for women." Warming to the subject, Friedan continued: "It is a terrifying denial of what should be basic rights for women, an utter denial of the personhood of women—the complete dehumanization of women. It is an important human rights case. To put it at the level of contract law is to dehumanize women and the human bond between mother and child."[26] Feminists homed in on Judge Sorkow's attack on Mary Beth Whitehead's competency to mother and on the degradations of the commercialization of surrogacy, with intermediaries receiving fees of $10,000 or more for arranging contracts; Friedan and others called them pimps. (Of course, this makes surrogate mothers prostitutes, but perhaps it is best not to spell this out.)

Some feminists pointed to the fine points in the contract that required that Mary Beth Whitehead abort on William Stern's demand should the fetus show any signs of "physiological abnor-

25. *New York Times,* 24 Mar. 1987, p. 26. Lipset, an assistant professor of anthropology, was pretty clearly out of his depth on this one.

26. Friedan, quoted in James Barron, "Views on Surrogacy Harden after Baby M Ruling," *New York Times,* 2 Apr. 1987, p. 16. Of course, at one point Friedan and others lodged the denial of the personhood of women in the assumption that a necessary bond existed linking biological and social motherhood; how ideological worms turn!

mality" following amniocentesis. Many found this repugnant and, yes, immoral—because the male got to order it, not because such abortion is repugnant on principle. All feminists aroused by this question circled around a vital point—that, in Friedan's words, "the claim of the woman who has carried the baby for nine months should take precedence over the claim of the man who has donated one of his 50 million sperm." The most eloquent statement of feminist outrage came from Katha Pollitt, who wrote in the *Nation*, "What William Stern wanted, however, was not just a perfect baby; the Sterns did not, in fact, seriously investigate adoption. He wanted a perfect baby with his genes and a medically vetted mother who would get out of his life forever immediately after giving birth."[27]

In the year or more since the initial shock of Baby M, feminist queasiness about feminist queasiness has begun to appear. Ellen Willis, a hard-line pro-choicer, put it this way in the *Village Voice:*

> To permit surrogacy for pay encourages the exploitation of women; to ban it limits women's autonomy. To let the biological mother abrogate a surrogacy contract if she changes her mind—and not give the father the same right—violates the concept of equal rights and responsibilities for both sexes; to force a woman to give up a child she has borne— or let a man decide not to take the child—denies the difference between pregnancy and sperm donation, between men's power and women's vulnerability.

Willis finds a way out—feminists just haven't been radical enough, and hence the following quandary (posed within the framework of ultra-liberal assumptions, of course, as Willis parses the matter): "Since it's only in a system in which children belong, literally, to their parents that the concept of surrogacy makes sense, the starting point for sorting out its contradictions should be questioning the nature of the family."[28]

We have come back, full circle, to concerns with the nature

27. Pollitt, "The Strange Case of Baby M," *Nation*, 23 May 1987, p. 688. Pollitt even had a few good words to say about the Vatican instruction on reproductive technology. The Vatican was right, she claimed, in insisting that you "don't have a right to a child, any more than you have a right to a spouse. You only have the right to try to have one."

28. Willis, "1987. Feminiso," *Village Voice*, 5 Jan. 1988, p. 21.

of human intimacy and the family. That is as it should be. The new eugenics cannot be disarticulated from a wider cultural and social surround. All eugenics worldviews with which I am familiar—from Plato's *Republic* to Hitler's Reich—aimed to eliminate, undermine, or leapfrog over the family in order to achieve their aims. The same holds for modern eugenicists: the family is a drag on radical forms of social and genetic engineering. Women's attachment to their own children is a problem. The fact that people continue to sort themselves out into family units is a problem. It would be far easier if natural pregnancy could somehow be phased out. But, in the meantime, newer and better ways to convince people to participate in eugenics efforts (under other names, of course) must be devised.

What makes this so complicated a concern for so many feminists is the fact that many spokesmen and women committed to a strong normative vision of the two-parent family, particularly those who sustain this vision with religious commitment, oppose the new eugenics. Feminists indebted to ultra-liberalism, having located the family as the root of oppression, find it difficult to make common cause with those who see the family as the preferred social site where vulnerable humanity should be nurtured and sustained. Thus one finds feminist attacks on the Warnock Report (the report of the Committee of Inquiry into Human Fertilisation and Embryology, requested by the government of Great Britain), because it states, "We believe that as a general rule it is better for children to be born into a two-parent family, with both father and mother." This gets construed as elitist and denying to women the possibility for "independent motherhood."[29]

My conclusion to these reflections, then, is a recognition: feminist quandaries concerning the new eugenics inexorably pitch feminists back into discussions of men, women, children, families, and the wider community. To insist, as all radical feminists and many equal-rights and Marxist feminists have done, that we must have done with evoking any norm of the family that consists of men, women, and offspring and must instead remain neutral about how people organize their private lives—this insis-

29. Patricia Spallone, "The Warnock Report: The Politics of Reproductive Technology," *Women's Studies International Forum* 9 (1986): 544.

tence may, paradoxically, have opened up those lives, including the lives of women, to more extensive forms of control. The evocations of "independent motherhood" that I kept running across in my readings are really rather sad. For the miniscule number who opt for such robust independence—whether as a partner in a lesbian relationship or as an economically well-off career woman who can afford full-time child care after she has given birth to her out-of-wedlock baby—there are tens of thousands of women who are displaced homemakers or client-dependents of an intrusive welfare-state bureaucracy. These women would *prefer* to be part of an intact family. Outside of that network, they are more, not less, vulnerable to the ministrations of experts, therapists, social workers—the lot.

But simply reaffirming a family norm will not do, either. Somehow we must find a compelling way to think about how the lives of all Americans (but especially the upper-middle-class, educated sort of Americans), male and female, are entangled with the atomist project, including the notion that one can and should achieve as much control as possible over one's life, including fertility. The search for intrusive intervention in human reproduction comes from those able to command the resources of the genetic engineers and medical reproduction experts. They are prepared to accept a remarkable degree of surveillance and manipulation of their intimate lives, given this fetish of control fused with the strange *demand* that babies can and must be made whenever the *want* is there. In this way human procreation is transformed into a technical operation, and that, in the long run, promotes a project of what Oliver O'Donovan calls "scientific self-transcendence." O'Donovan claims that in our own culture *curiositas* has become a "sin of the masses. . . . The liberal revolution arose, and will continue to evolve, in answer to a mass desire of Western civilization, in which we all participate and not at the behest of a few scientists."[30] It is important to locate feminism within this wider project and to recognize that many feminists are troubled by the Frankenstein we seem to be unleashing.

30. O'Donovan, *Begotten or Made? Human Procreation and Medical Technique* (Oxford: Oxford University Press, 1984), p. 9.

The Child as Chattel: Reflections on a Vatican Document

James Tunstead Burtchaell, C.S.C.

The Vatican's Congregation for the Doctrine of the Faith (CDF) aroused a firestorm of criticism with its recent *Instruction on Respect for Human Life in Its Origin and on the Dignity of Procreation: Replies to Certain Questions of the Day.*[1] U.S. theologians promptly began to call the Instruction "unpersuasive."[2] Various Catholic university hospitals in Europe announced that they were going to ignore it and continue their fertility services.[3] The *New York Times* found it an occasion to reiterate its doctrine concerning official Catholic authorities: like the earlier bans on abortion and contraception, this condemnation was an act of resistance against enlightenment, against a technology which the editors regard as "an enormous step into the future."[4]

The document has been heavily discounted because it deals with procreation and was presumably written by male celibates

---

1. *Origins,* 16, 40 (19 Mar. 1987). Subsequent references to this source will abbreviate the title as *Instruction.* For response, see the *New York Times,* 10 Mar. 1987, p. Y10.

2. *New York Times,* 11 Mar. 1987, p. 1.

3. *New York Times,* 18 Mar. 1987, p. A1.

4. *New York Times,* 12 Mar. 1987, p. A30.

who must lack an insider's knowledge.[5] Lay dismay was widely cited as well. A forty-year-old American woman who had adopted one child and with her husband was trying to conceive through in vitro fertilization (IVF) said, "Parenting is such a strong urge. I don't think the church can stop it."[6] An Australian wife who hopes for her second child via the same technology, complains, "All we're asking for is to allow a man and wife to have a family."[7]

The document, which was published over the signature of the prefect of the CDF, Joseph Cardinal Ratzinger (with the annotation that Pope John Paul had personally authorized it), does set forth a rationale for its ethical conclusions. That has not always been the style of this Roman agency. In 1679, when it bore the more ominous title of Sacred Congregation of the Universal Inquisition, it issued—without any supporting explanations—a catalogue of assorted errors on matters moral. That list included condemned propositions such as these:

> Eating and drinking one's fill for the sheer pleasure of it is not sinful, provided it creates no health risk, because any natural appetite can licitly find enjoyment in its own pursuits.

> It is no fault—indeed, not even a minor shortcoming—for a couple to join in marital intercourse for the sheer pleasure of it.[8]

Innocent XI, the pope at that time, added to this list of sixty-five censured teachings a postscript that warned contentious academics not to be too free in condemning one another's theories. They were to refrain from making accusations (at least against their fellow Catholics) and leave the task of condemnation to the Holy See.[9]

In 1907, just before Pius X gave the agency a new name—the Sacred Congregation of the Holy Office—another blacklist

5. Daniel C. Maguire, "Vatican Birth Decree Casts a Long Shadow," *Boston Globe*, 15 Mar. 1987, pp. A21-24.

6. Cited in *New York Times*, 12 Mar. 1987, p. B11.

7. Cited in *New York Times*, 19 Mar. 1987, p. Y6.

8. Decree of 4 March 1679, in Henricus Denzinger et al., *Enchiridion Symbolorum*, nos. 1158, 1159.

9. *Enchiridion Symbolorum*, no. 1216.

was published. Coincidentally, it also included sixty-five propositions, including these:

> The Church shows herself incompetent to promote
> the gospel ethic effectively because she stubbornly clings to
> unchangeable doctrines which cannot be accommodated to
> modern progress.

> Contemporary Catholicism cannot keep in step with
> sound science unless it is refashioned into a nondogmatic
> Christianity, like a generic and liberal Protestantism.[10]

CDF, the current legatee of that terse and nay-saying prose style,
is charged with expounding and clarifying Christian faith. It must
beat the bounds of the church and tend the fences that mark the
borders of sound belief. If this recent publication, despite its efforts to give out the whys and wherefores, should fail to be persuasive even among many thoughtful Catholics, that may be due
to certain systemic handicaps within the CDF itself.

Among the twenty cardinals and bishops listed as official
members, less than a handful are known to have been professionally trained and experienced as practicing theologians. The
in-house staff, numbering hardly a dozen persons, lists no one
known as a leader among scholars. There is a board of academic
consultors, among whom faculty members at local Roman sem-
inaries predominate.

The task of scrutinizing new attempts to interpret and apply
the doctrine of a community scattered across all the cultures of
the world, with an intellectual lineage going back to the apos-
tles—indeed, back to Abraham—is surely a work for skilled intellectuals. It would seem that an agency charged with that
scrutiny ought not be denied the resources needed to produce
perceptive and cogent teaching on complex and crucial issues.
Full many a theologian is born to blush in Vatican offices unseen,
but such an accumulation of anonymity as this virtually guarantees that the CDF never will have the capacity to reply effectively to "certain questions of the day."

Nevertheless, I want to argue that the insight which the Instruction so stiffly represents and unpersuasively defends is both
true and timely.

10. Decree of 3 July 1907, *Enchiridion Symbolorum*, nos. 2063, 2065.

The Vatican is well aware that its adversary in this matter of human procreation is modern technology. Now the technologist is a "can do" person whose success is usually proportionate to his ability to fix total attention upon a given problem, to isolate it from all extrinsic factors, and resolutely to manipulate the variables in order to produce the client's satisfaction. The technologist tends to see into a problem by squinting, by narrowing the gaze.

Only a few days after the Instruction appeared, I had occasion to admire the pluck and pertinacity of modern technology near Boston, where I was visiting friends. The affluent townships encircling that city lodge the executives of the corporate world which assembles each morning in the downtown business district. The problem presented to Boston technology was how to get those suburban executives to work and back with a minimum of inconvenience. The solution was to build the Massachusetts Turnpike straight out and through Weston. And, so that the commuters would not have to slow their pace during the snow and ice of winter, hundreds of tons of industrial salts have been strewn upon the turnpike's surface each year. The result is—as specified—that the good residents of Weston can be at their desks after a brisk and direct journey in the morning, and back out amid the leafy wonderland of ponds and copses and meadows in the evening.

But there have been additional results. The gracefully contoured countryside of the area, which is one of the most moneyed neighborhoods in the United States, has had the settled life of its wetlands deranged, and its water runoff is now disturbed. The water of the town reservoir has been rendered saltier than the Atlantic Ocean and is undrinkable.

The vulnerability of contemporary technology consists in its readiness to view tasks no more largely than the desires they are asked to satisfy. George Parkin Grant has written, "Thought is steadfast attention to the whole."[11] I would take that to be a pretty good description of classical Christian morality as well. The fellows who built the turnpike did just what the burgesses beyond Boston wanted. And they ignored the fact that nature cannot be refashioned on demand. They steadfastly ignored the whole to serve the interests of a segment of the whole. And that is what sin is.

11. Grant, *English-Speaking Justice*, ed. Stanley Hauerwas (Notre Dame, Ind.: University of Notre Dame Press, 1985), p. 87.

Look southward for another moral tale. The Ogallala Aquifer is possibly the largest underground freshwater source in the world. Eight Great Plains states depend upon it for their water. The aquifer was created of sand and gravel deposits more than a million years ago. It fills slowly; about 700 years of rainwater seepage is needed to replenish it.

When farmers in the Southwest recently decided to augment their agricultural output and financial income, they began drawing on that water supply more thirstily. They now extract nearly ten times as much water as the aquifer can recapture each year. The water table sinks lower each month. The subsoil is consequently collapsing and becoming too compacted to take in the water needed to recharge the reservoir below. As in Weston, there is a twofold result. Increased irrigation has increased agricultural production and profit, exactly as desired. But by failing to defer to nature, the users have guaranteed that virtually all farmland from Kansas to Texas, from Arizona to Arkansas will be wasteland in thirty or forty years. It is not that the users act in ignorance: they know the facts. But they avert their gaze. There is no steadfast attention to the whole. In its willful and impetuous determination to fit nature to its will, technology is irreversibly violating the earth. The present dwellers have impoverished their offspring. That is a sin.[12]

Twenty-five years ago the *New Yorker* first published Rachel Carson's *Silent Spring*, an exposé of pesticides, which she called the "elixirs of death." The sophisticated and very modern chemicals that technology had produced on request to control insects and other crop predators were irreversibly poisoning the earth and the air and the water and the living organisms with which humans must abide in respectful partnership. Carson inscribed her book with Albert Schweitzer's complaint: "Man has lost the capacity to foresee and to forestall. He will end by destroying the earth."

We must cultivate, Carson argued,

> the awareness that we are dealing with life—with living populations and all their pressures and counter-pressures, their surges and recessions. Only by taking account of such

---

12. See "The Browning of America," *Newsweek*, 23 Feb. 1981, pp. 26-37; "Water: The Nation's Next Resource Crisis," *U.S. News and World Report*, 18 Mar. 1985, pp. 64-68.

life forces and by cautiously seeking to guide them into channels favorable to ourselves can we hope to achieve a reasonable accommodation between the insect hordes and ourselves.

The current vogue for poisons has failed utterly to take into account these most fundamental considerations. As crude a weapon as the cave man's club, the chemical barrage has been hurled against the fabric of life—a fabric on the one hand delicate and destructible, on the other miraculously tough and resilient, and capable of striking back in unexpected ways. These extraordinary capacities of life have been ignored by the practitioners of chemical control who have brought to their task no "high-minded orientation," no humility before the vast forces with which they tamper.

The "control of nature" is a phrase conceived in arrogance, born of the Neanderthal age of biology and philosophy, when it was supposed that nature exists for the convenience of man. The concepts and practices of applied entomology for the most part date from that Stone Age of science. It is our alarming misfortune that so primitive a science has armed itself with the most modern and terrible weapons, and that in turning them against the insects it has also turned them against the earth.[13]

The technologists were not amused to be told they were using modern methods but primitive attitudes. They denounced Carson as shrill, hysterical, romantic. The most cutting denunciation was that she was backward, that she was flailing at technological progress.[14] But the tides eventually pulled with Miss Carson, and her book was one of the inaugural works of the environmental movement, born of her "anger at the senseless, brutish things that were being done."[15]

It is not technology itself that is sinister, nor its method, which is to restrict its gaze to the problem at hand. The damage comes

13. Carson, *Silent Spring* (Boston: Houghton Mifflin, 1962), pp. 296-97.

14. Norman Boucher, "The Legacy of *Silent Spring*," *Boston Globe Magazine*, 15 Mar. 1987, pp. 17ff.

15. Paul Brooks, *Speaking for Nature* (Boston: Houghton Mifflin, 1980), p. 286.

when a technologist (or any professional) acts only on the competence of his or her discipline while blind to any wider view of the fuller human agenda. The technologist goes berserk and and becomes the sorcerer's apprentice when she devises tactics in ignorance of what strategy or policy or ethic they are going to serve.

The ecological movement, now well underway, is the first effectively persuasive rebuke to a belief that had grown apace with modern science and technology since the Enlightenment. The doctrine held that human purpose and choice would encounter no limits in their dominion over nature. Or, to put a finer point on the doctrine, it held that there was no nature, no creation with an innate character and requirements quite beyond what we could choose to infuse in things or bleach out of them.

But the environmental disillusionment had to do with the elements and with plants and animals. We have yet to be persuaded that there are given natural forces and needs in humanity itself that may be as aloof from our willful choices as are the ocean tides that make sport of our retaining walls and waterways.

We are more conscious now that a healthy diet of bread and legumes and fish and fresh vegetables and fruits is naturally good and is hardly enhanced by lashings of sodium hexametaphosphate, FD & C Yellow No. 5, monosodium glutamate, or potassium sorbate. And we sense that a sedentary pace coupled with gluttony at table must not be facilitated by stapling the stomach or excising half of the bowel or surgically stripping off a girdle of excess fat. And there is enough alarm at the readiness of some physicians to overmedicate that now some nurses will not take even an aspirin tablet to fight a headache.

This reticence toward technological intrusion for the "control of nature" is really a later application of ecological conscience to our own bodies. But if we go deeper, if our steadfast pursuit of the whole draws us beyond our physical selves to our distinctively human and personal makeup, are we prepared to doubt that we can be whoever or whatever we choose, unconstrained by what we were created and fashioned to be?

The movement to assist sexual intercourse with technology was begun by two valiant women: Marie Stopes in Great Britain and Margaret Sanger in the United States. The prescribed purposes of this movement were clear and cogent. Inferior members of the race were to be prevented from breeding, and worthwhile persons were to be spared the burden of begetting more children than they wanted to care for.

The eugenic agenda was explained by Dr. Stopes:

> It is perhaps only to be expected that the more conscientious, the more thrifty, and the more lovingly desirous to do the best for their children people are, the more do they restrict their families, in the interests both of the children they have and of the community which would otherwise be burdened by their offspring did they not themselves adequately provide for them. Those who are less conscientious, less full of forethought, and less able to provide for the children they bear, and more willing to accept public aid directly and indirectly, are more reckless in the production of large families. . . .
>
> The thriftless who breed so rapidly tend by that very fact to bring forth children who are weakened and handicapped by physical as well as mental warping and disease, and at the same time to demand their support from the sound and thrifty.[16]

For the very reason that worthwhile people were conscientious and the undesirables were not, contraception was the technology proposed for the "fit." Something less voluntary was appropriate for the unfit. "It seems easy enough to supply the intelligent and careful woman with physiological help; and for the careless, stupid or feeble-minded who persist in producing infants of no value to the State and often only a charge upon it, the right course seems to be sterilization."[17]

Mrs. Sanger advocated an even more vigorous application of remedies:

> [The first step would be] to apply a stern and rigid policy of sterilization to that grade of population whose progeny is already tainted [the feebleminded, idiots, morons, the insane, the syphilitic, the epileptic, the criminal, professional prostitutes, etc.], or whose inheritance is such that objec-

16. Stopes, *Wise Parenthood*, 9th ed. (London: Putnam, 1922), pp. 17-19.

17. Stopes, *Wise Parenthood*, p. 44.

tionable traits may be transmitted to offspring . . . also, to give certain dysgenic groups in our population their choice of sterilization or segregation.

The second step would be to take an inventory of the secondary group such as illiterates, paupers, unemployables, criminals, prostitutes, dope-fiends; classify them in special departments under government medical protection, and segregate them on farms and open spaces as long as necessary for the strengthening and development of moral conduct.[18] [She expected that the first group might comprise five million people; the second, possibly twenty million.]

Historian Daniel J. Kevles recounts the early orientation of this movement: "Before the war, Sanger had linked birth control with feminism. Now, like her British counterpart Marie Stopes, she tied contraception increasingly to the eugenic cause. 'More children from the fit, less from the unfit: that is the chief issue of birth control,' said Sanger."[19] Once sterilization was used to prevent defective and morally objectionable people from further tainting the future citizenry, good people of modest means could be provided with safe and healthy contraceptives. This dual program of fertility control would obviate two great horrors: abortion and war.

There was an early recognition that this progressive undertaking would go down badly with Catholic officials. Stopes wrote,

> The Roman Catholic Church in particular is the most unyielding in its total condemnation of the use of scientific aid in controlling the production of children, although it—like the other Churches—concedes the *principle* of the justifiability of control in some circumstances. To concede the principle, even while condemning the best methods of effecting such control, is to deny the uses of intellectual progress.[20]

There would be others disposed to stand in the way of such a plucky, "can do" venture, Stopes noted, and she offered this

18. Sanger, "A Plan for Peace," *Birth Control Review* 16 (Apr. 1932): 107-8.
19. Kevles, *In the Name of Eugenics: Genetics and the Uses of Human Heredity* (New York: Knopf, 1985), p. 90.
20. Stopes, *Wise Parenthood*, p. 13.

refutation: "To those who protest that we have no right to interfere with the course of Nature, one must point out that the whole
of civilization, everything which separates men from animals, is
an interference with what such people commonly call Nature."[21]

The venture succeeded. Sterilization and contraception are
now legal throughout the land, endorsed by virtually all religious
bodies, advertised openly and urged with subsidy upon the poor.
There are, in fact, more sterilizations performed annually upon
Americans today than there were American marriages being performed when Stopes and Sanger began their work of enlightenment.

But the venture of technology overshot its original goals
in several unforeseen ways. America now has the highest rate
of teenage pregnancy in what we like to call the developed
world. The incidence of sexual promiscuity, venereal disease,
abortion, marital collapse, fatherless children, child abuse and
abandonment, and wanton breeding by the thriftless (and
dusky) poor is probably higher than at any other time in our national history.

The founders did concede that their program might not immediately achieve its desired results. But that was because of the
lack, in their time, of an "ideal" contraceptive. Its development
was to be expected from further assiduous scientific research.
Seventy years and much assiduous research later, the same message is still bravely recited. With more technology—more instruction in contraception, more availability of contraceptives, and
more access to abortion as a fail-safe backup—all of these
problems can be licked. There is nothing that enlightenment and
scientific progress cannot overcome.

Planned Parenthood and its research organ, the Alan Guttmacher Institute (AGI), are aware that their accomplishments still
fall short of their hopes. Yet an enormous breakthrough has recently put success within their reach at last. The problem had previously
been viewed too narrowly, in terms too exclusively biological.

Some researchers have begun to attack the problem of
teenage pregnancy from a sociological perspective. Their
studies suggest that if the problem arises mainly out of cul-

21. Stopes, *Wise Parenthood*, pp. 24-25.

tural and socio-economic conditions, the longer-term solutions may also have to address those same conditions.

The evolution of Joy Dryfoos' own thinking in this regard is instructive. Dryfoos, one of the country's leading experts on the issue of school-age pregnancy and former planning director of the AGI, says: "One of the last things I did at the Guttmacher Institute was to help prepare *Teenage Pregnancy: The Problem That Hasn't Gone Away*. It was a statistical pamphlet that documented the need for sex education, access to contraception and abortion services, and services for adolescent parents and their children. And yet I knew that, even doing all those things, the issue of teenage pregnancy would not go away. And it wouldn't go away because there are young people who do not perceive that there is any reason to use contraceptives and avoid parenthood."[22]

The problem, then, is motivation: not motivation regarding sex or friendship or love or childbearing but motivation to use technology when one refuses to think of those other—less simple and more troubling—matters. So new programs to encourage the use of contraceptives are now underway.

On other fronts, where motivation is less sluggish, the biotechnology has been producing admirable results. For those who beget children but do not desire them, pharmaceutical preparations so safe that one can keep them in a medicine cabinet and dose oneself at home will regularly expel the product of conception. For those who beget children but want to verify whether or not they are desirable, amniocentesis or chorionic villi sampling can examine the unborn for defects before it is too late. For those who beget undesirable children but fail to determine that and destroy them before birth, compassionate doctors have devised an acceptable medical way to minister to their chagrin by orders to withhold all nourishment and liquids until it is assured that parental desires will be satisfied. For those who desire children but fail to beget them, in vitro fertilization (IVF) occasionally gives them their heart's desire. For those who desire children but not pregnancy, other

22. "Adolescent Pregnancy: Testing Prevention Strategies," *Carnegie Quarterly* 31 (Summer/Fall 1986): 4.

women have been put at their disposal to breed for them: your egg or hers, as specified.

In the face of such awesome, versatile, abundantly funded accomplishment, it appears never to have occurred to the authors of the technical assistance that their very program might, for all its proficiency of method, be "senseless and brutish" in its imagination and results. The naiveté of the technicians seems never to have been jolted by any suspicion that their massive scheme of helpful contrivances might itself have been one of the surest stimulants of irresponsible sexual activity. It escaped them that their efforts might have been to sex and childbearing what the turnpike and salt have been to the Weston groundwaters and reservoir.

One is reminded somewhat of the editorial amazement expressed in our land when the voters in Ireland were presented recently with a measure that would finally legalize divorce and remarriage. Stories were told of Irish couples whose second or third unions were suffering for want of legal recognition. It was, we were told, such a backward thing for Ireland to deprive these good folks of a remedy for their problems. Here were the literati of America, with runaway rates of adultery, divorce, unsuccessful second and subsequent marriages, lecturing the Irish on how they would improve their domestic relations if only they would adopt our custom of divorce as the remedy for marital breakdown.

To have addressed the problems of undesired pregnancy without ever considering that it is the sexual irresponsibility itself which needs to be confronted; to remedy the chronic failure of a program composed entirely of contraception, sterilization, and abortion by providing nothing but more contraception, sterilization, and abortion—this can only remind one of the claim by American military commanders in Vietnam that if only they could have a few hundred thousand more troops, and more cluster bombs and fragmentation bombs and Agent Orange and all that other good stuff which wins wars, then the campaign could be won.

Meanwhile, back at the Vatican, the pope and his theological subalterns have borne high annoyance, frustrated because they have been so unsuccessful in rousing us all to alarm. Among his scriveners, Cardinal Ratzinger obviously employs no one like

Rachel Carson. They all write with a grace and a clarity that are not user-friendly.

An attentive reader of the *Instruction on Respect for Human Life in Its Origin and on the Dignity of Procreation: Replies to Certain Questions of the Day* in its official English version, however, encounters some lively new vocabulary. One expects mention of natural law, and it does make an appearance here. Eight times the text speaks of the natural moral law, a divine law that is unwritten in any sacred text but inscribed in the very being of men and women and their orientation to sexual union. But this is a minor motif. Much more prominent in the argument is an insistence on "dignity," on "rights," and on "respect" for human dignity and rights. Together these terms appear twelve times more frequently than all references to moral law.

From the first paragraph on, the Instruction repeatedly asserts the dignity of the human person, of spouses, of parents, of children. The origin of humankind, procreation through marital union, is also possessed of dignity. When the talk shifts to rights, the document reaches for emphatic adjectives. It speaks of fundamental rights, inviolable and inalienable rights. The right of every innocent human to life itself is primary. Next comes the right of every person to be respected for oneself, and also to be assured in marriage of exclusive sexual intimacy with one's spouse, and the right to beget any children with that spouse alone. Several times the document speaks of a right that is not familiar from previous Vatican statements: the right to be conceived, carried in the womb, welcomed into the world, and brought up within marriage.[23] "It is through the secure and recognized relationship to his own parents that the child can discover his own identity and achieve his own proper human development."[24] Near the center of its argument, however, there is one very fundamental right that the Instruction denies—the right to possess a child:

> A true and proper right to a child would be contrary to the child's dignity and nature. The child is not an object to which one has a right nor can he [or she] be considered as an object of ownership. Rather, a child is a gift, "the supreme gift" and

23. *Instruction,* II.6.a.
24. *Instruction,* II.1.c.

the most gratuitous gift of marriage, and is a living testimony of the mutual giving of his [or her] parents.[25]

All humans are called to a respect for each other that is unconditional and absolute. The child, no matter how young, must be respected as one's equal in human dignity from the moment of conception. In his conclusion Cardinal Ratzinger characterizes the entire Instruction as an appeal for respect, in family and in society, for human life and human love. The dark opposite of such a respect is contempt for both. "The Congregation . . . hopes that all will understand the incompatibility between recognition of the dignity of the human person and contempt for life and love, between faith in the living God and the claim to decide arbitrarily the origin and fate of a human being."[26]

Pope John Paul II himself is responsible for this newly preferred vocabulary. It goes some way toward talking the language of contemporary social advocacy movements. Despite the soothing sound of "rights" and "dignity" to the establishment liberals that direct most of those movements, their dogmas do not bend sufficiently to allow many of them much sympathy with this part of his agenda. The words are the words of Voltaire and Tom Paine, but the thoughts are the thoughts of Jesus.

Even with this venture into contemporary idiom, the Roman text labors. This must be due in part to the Vatican's abhorrence of rhetorical flair or eloquence. Still, one asks why the prose is so curt, why it states its most contentious and confrontational and prophetic insights so flatly that they have little chance to engage the reader's mind or imagination. When two parties have shared considerable experience and long conversation to the point where they possess a common recollection and interpretation, they can evoke that common understanding later by a brief word or phrase, even a glance across a crowded room. But to any outsider who has never participated in the exploring or the resolving of the same matter, that word or phrase would have no ring or sense to it. The allusions would be missing.

Here we are listening to the man assigned the most venerable rostrum in the Catholic Church in order to address some of

25. *Instruction,* II.8.c; gender inclusion added.
26. *Instruction,* conclusion, par. 2.

the most profound life-and-death issues over which the vision of faith is at war with the public culture. It simply will not do for him to state those convictions tersely as if most readers—indeed, even most Catholic readers—had cared, and thought them through, and concurred. He must plead his points, and illustrate his convictions, and win our minds and imaginations to his commitment.

Consider one of the themes most central to the Instruction: that biology and medicine can heal, but the powers they deploy can as easily be used to injure or destroy people. The point is stated, but never offered enough chance to gather strength:

> Medicine which seeks to be ordered to the integral good of the person must respect the specifically human values of sexuality. The doctor is at the service of persons and of human procreation. He does not have the authority to dispose of them or to decide their fate. . . . The humanization of medicine, which is insisted upon today by everyone, requires respect for the integral dignity of the human person first of all in the act and at the moment in which the spouses transmit life to a new person.[27]

> If the legislator responsible for the common good were not watchful, he could be deprived of his prerogatives by researchers claiming to govern humanity in the name of the biological discoveries and the alleged "improvement" processes which they would draw from those discoveries. "Eugenism" and forms of discrimination between human beings could come to be legitimized: This would constitute an act of violence and a serious offense to the equality, dignity and fundamental rights of the human person.[28]

These are potent ideas, strenuous worries. Rome sees itself defending humanity from the excesses of our own power,[29] yet it speaks with a rhetoric that could keep only a philosopher awake. The point is stated, but it is not propounded.

The legitimate role of technology to assist in human procreation must be vigilantly supervised, for it is as ready to overpower humans as it is to be of help to them:

27. *Instruction*, II.7.b, d.
28. *Instruction*, III, par. 2.
29. *Instruction*, conclusion, par. 3.

Thanks to the progress of the biological and medical sciences, man has at his disposal ever more effective therapeutic resources; but he can also acquire new powers, with unforeseeable consequences, over human life at its very beginning and in its first stages. Various procedures now make it possible to intervene not only in order to assist, but also to dominate the processes of procreation. These techniques can enable man to "take in hand his own destiny," but they also expose him "to the temptation to go beyond the limits of a reasonable dominion over nature."[30]

Through these procedures [IVF, which creates new human persons and then discards some of them], with apparently contrary purposes, life and death are subjected to the decision of man, who thus sets himself up as the giver of life and death by decree. This dynamic of violence and domination may remain unnoticed by those very individuals who, in wishing to utilize this procedure, become subject to it themselves. The facts recorded and the cold logic which links them must be taken into consideration for a moral judgment on in vitro fertilization and embryo transfer: The abortion mentality which has made this procedure possible thus leads, whether one wants it or not, to man's domination over the life and death of his fellow human beings and can lead to a system of radical eugenics.[31]

Technology *macht frei!* This is Rachel Carson stuff, only escalated from sedge and bracken, plovers and thrushes to men and women and girls and boys. The same themes are here. Technology is invited in to alleviate some of the ills of life, and it can, if given its own head, seize control of the very lives it was asked to assist, with no one the wiser about what had happened.

Another scenario portrays those who are involved in technology not as its victims but as its users, exploiting others. The Instruction pursues this latter point when it warns parents of their temptation and ability to treat their own offspring without the respect due them as human equals:

The human person must be accepted in his [or her] parents' act of union and love; the generation of a child must therefore be the fruit of that mutual giving which is realized

30. *Instruction,* introduction, 1, par. 2.
31. *Instruction,* II, par. 3.

in the conjugal act wherein the spouses cooperate as servants and not as masters in the work of the Creator, who is
love. . . . He [or she] cannot be desired or conceived as the
product of medical or biological techniques; that would be
equivalent to reducing him [or her] to an object of scientific
technology.[32]

The Vatican is too technical, or perhaps too dainty, to state
graphically enough that we have been turning procreation into
science fiction, and that we become monsters as a result. A society
that venerates Drs. Masters and Johnson and their lab-coat lore of
orgasm as advisors on the fullness of human sexuality, or that
hearkens to Dr. Ruth as a *sage femme* of how men and women give
themselves to one another, or that orders up children the same
way it uses the Lands' End catalogue—this is a creature-feature
that ought not appear even on late-night Saturday television. Or
so I take the Vatican to be telling us.

Turn now to the second central tenet of the Instruction: the
demand within human nature that sexual intercourse be a vital
and intrinsic link not only between spouses but also between
parents and children. The Catholic understanding of human
procreation is bewilderingly compact, and it begins something
like this. It is in a committed and steadfast community that we
find the vision and the gumption to emerge from our infantile sel-
fishness into mature love. The two closest bonds that sustain us
through life are the enduring devotion we receive from our
parents and then give to our children, and the enduring devotion
we give and receive by pledge with our spouses: the bonds of
blood and the bonds of promise. These commitments are life-
giving, Catholics observe, only if both of them are unconditional
and lifelong: for better or for worse, until death. And the two
bonds find their ligature in sexual intercourse, wherein the embrace of two in one life gives flesh-fruit in children. This is what
lies behind the following statement in the Instruction:

> The church's teaching on marriage and human procrea-
> tion affirms the "inseparable connection, willed by God
> and unable to be broken by [humans] on [their] own initia-
> tive, between the two meanings of the conjugal act: the
> unitive meaning and the procreative meaning. Indeed, by

32. *Instruction,* II.4.c.1, 2; gender inclusion added.

its intimate structure the conjugal act, while most closely uniting husband and wife, capacitates them for the generation of new lives according to laws inscribed in the very being of man and of woman" [*Humanae Vitae*]. This principle, which is based upon the nature of marriage and the intimate connection of the goods of marriage, has well-known consequences on the level of responsible fatherhood and motherhood.[33]

The insight that the Vatican has been honoring more obstinately than persuasively through the years is that it is both naturally necessary and, by God's grace, possible for human persons to celebrate *both* of their most loyal commitments in sex. That is where a husband and wife are most vividly and revealingly summoned into fidelity, which is one of our better ventures at imitating the love God has for us. He loves and embraces us, regardless of what we have done or could do ourselves, out of the constancy of his benevolence. It is very difficult for us to grow into such an unconditional love. The marital union is a unique schooling in that love, in that a woman or a man can clasp close a partner not because of that other person's qualities or accomplishments. And in accepting one another on those terms, they accept their children on those same terms, children who quite literally issue forth from that union.

From the Catholic reflection on experience, this is how we have to live. To others who find it too demanding and call it an ideal, the church replies that it is an ideal like drinkable water is an ideal, like non-carcinogenic food is an ideal, like cars that are not unsafe at any speed are an ideal, like telling the truth is an ideal.

From such a perspective, it is no surprise that contempt for marital fidelity would provoke a disintegration of the sexual experience, and that this in turn would threaten both the welfare and the very existence of children. The erotic is divorced from the relational, and both are estranged from the procreative. All aspects alike will wither by their separation.

If man and woman do not pledge themselves for better or for worse until death, then their relationship is wary and hedged and revocable. Choice becomes the predator of commitment. In

33. *Instruction*, II.4.a.1; gender inclusion added.

such circumstances children cannot be welcomed into an assured embrace, for they too are now subjected to scrutiny and conditional loyalty. They are similarly victims of the defeat of commitment by unlimited freedom of choice. Their security is hedged because the fidelity from their parents is also revocable. What person could grow in that sterile a setting?

Rome, which is haunted by the conviction that commitment entails the forfeiture of some future free choices and that without fidelity we must wither, wails to the world that we cannot—we dare not, we must not—have the arrogance to make every relationship in our lives revocable and still expect to survive.

Rome is invoking the old argument from nature. We are simply not made to survive under certain conditions, and it is up to a perceptive community of faith to draw upon its cumulative experience and to formulate its observations of what such conditions usually are. When the Instruction does invoke the old expression "natural law," it means what Rachel Carson meant: that we cannot manipulate our cosmos or our society or our persons or our bodies at will and still survive.

An era so appreciative of bran and ozone and Lamaze cannot be all that offended by insistence upon what is "natural." Perhaps it is the word "law" that most offends us. What law connotes is somebody else's will coercing mine. For a generation whose velocity on the highways and whose creativity in drawing up tax returns imply that the law has bite to it only if we are apprehended and punished, "law" may be a very deceptive metaphor for nature. For when you cheat on nature in a serious and chronic way, you are enfeebled, you sicken, you die. It makes no essential difference whether you are caught in the act or not. You suffer. Whether or not your doctor finds out you haven't quit chain-smoking, the emphysema is likely to ensue. And if your deviance is more moral than physical, you can die spiritually yet still keep getting out of bed every morning.

Technology is art: complex and purposeful artifice. And art, as Burke has said, is itself what human nature is all about. But art, as we need no Burke to tell us, has the capacity either to enhance our natural lives or to afflict them.

Technology has been misunderstood because nature and art have been misunderstood. Nature has been considered to be what is determinate within a human being, what can be studied by physics, chemistry, and biology, or, as some would have it, by psychology. Artifice, according to this same view, is what is not determined but free in us: choice. It is the subject of economics, politics, and ethics. Once those two elements had been divvied up and allotted to their respective experts, a major task assigned to technology was to humanize the physical, animal aspects of our experience by subjecting them to our free preferences and choices.

This seemed at first to correct the naive, materialist way in which our "lower" and "higher" elements were being studied independently of one another. Human nutrition, coupling, reproduction, exertion, locomotion, recreation, combat, and death could never be rightly understood or undertaken if they were treated as merely animal events. Lactation, to take one example, is not something we simply share with other mammals. For humans it needs to be invested with love and self-sacrifice that transcend and transfigure any lamb's or whale's experience at the breast. Mammalian suckling is not a specimen or a prototype of what human mothers and children do together; it is but a metaphor of human nursing. For many of our physical actions are inspirited by our intellectual purposes and needs and relations.

The technological mission, then, could be construed as uplifting human nature by making all physical aspects of human life responsive to free choice, our highest faculty. But what was being ignored all along was the counterpart truth: that our powers of judgment and decision are those of embodied creatures, and they may not work their whim or will with our physical selves by acting like foreign occupation forces. This is manifest in our case in point—human reproduction—for the dual reason that it cannot be defined in narrowly physical terms, being an undertaking of body and spirit together, and that it has inherent purposes and vectors of its own quite resistant to some we might prefer. It requires the intense involvement of our spirit, but not merely by way of absolute mastery.

Our nature is mysteriously composite: the will is not quite a stranger nor yet quite a sovereign; the gastro-intestinal tract is not quite autonomous but also not quite controlled. It is our nature to expectorate, to calculate, and to defecate, just as it is our nature to marry, to chew, and to owe loyalty to our own people and governors. All are physical acts, and each is spiritual; all are subject to our governance, but it is that of a constitutional—not an absolute—rule.

The inherent needs and fulfillment of our lower functions are typically less ambiguous than of those more evolved functions. A healthy diet, for instance, allows of more exact science than a sound economy, though we naturally need them both. Vitality through exercise is easier to work out than stability in marriage—yet neither is merely physical, and neither is merely relational. Each requires the satisfaction of our inherent nature and the pursuit of our chosen purposes. Thus any enterprise like that of technology, which invites the control of our natural lives, must give steadfast attention to the whole.

And this is not easily done. My friends in engineering confess chronic impatience with legally required environmental-impact studies. No matter how painstaking the work required, for example, to design a power plant to serve intricate residential and commercial markets, it proceeds on winged feet compared with the studies and licensures which assure that all the other interests and needs and resources of the community and the future community are being honored. That is because power plants have a way of doing some things beyond what was originally intended. It takes so much less time to produce a part than to accommodate it to the whole.

In her doctrine of nature and artifice, and the resulting criticism of technology that is reckless and improvident, Rome is speaking with the cautionary voice of tradition, issuing, as it were, an interior environmental-impact statement. In the other large matter at hand—the relationship between intention and action—Rome and her adversaries may both be pulling the classic tradition off its course, in opposite directions.

The ancient conviction was that individual acts could, by divine enablement and through determination and discipline, form virtues. Likewise, by wastage of divine grace and by neglect and self-indulgence, individual acts form vices. The amalgam of habituated good or evil was seen to give shape to a person's character, to one's fundamental commitment: to oneself or to others. Acts have an inherent, structured dynamic that tends to shape character, just as telling lies tends to make one dishonest not so much by one's withholding the truth from others but by one's losing the gumption to face it oneself. Conversely, habits of character tend to invest individual acts with meaning. Thus the difference between child abuse and a salutary swat resides not in the dynamics of force but in the disposition of the batterer.

Catholic moral theology in the Middle Ages had lost sight of this dialectic whereby human behavior natively embodied both habits of character and the moral force of actions, each affecting the other. A shift to legalism gave sovereignty to acts sundered from their habits. According to this view, a single act, no matter how uncharacteristic of a person's development and character, might be morally lethal because it was grievously forbidden.

The Reformation and the Enlightenment retaliated, not just by withdrawing definitive moral meaning from actions but by ignoring an ability in acts to create and reinforce, or to enfeeble and destroy, character. The Reformers did this by belittling human agency in good acts. The Enlightenment did it by denying any inherent, repercussive power of acts upon the self, and by replacing habitual character with free choice.

Here, then, was the perfect standoff. The conservatives (if I may so characterize them) were arguing that a single action stood in its total significance without reference to the person's disposition or habitual behavior. The liberals were arguing that every act received its total significance from the disposition of the agent—only that disposition was not the (traditionally conceived) character forged by long and reinforced behavior but the momentary whim or wish known as free choice.

Bring those two opponents to the issue of reproductive technology, and their respective doctrines will be predictable. Rome has said that a single sexual act of intercourse without openness to conception or a single act of reproduction without sexual intercourse is wrong by its very nature. The folks in white lab coats are saying that the entitlement to pleasure and the right to children of choice authorize us to pursue our preferences by any variation upon sex and reproduction, in the confidence that our freedom of choice will invest that sex, however conceivable or inconceivable, with our good intentions.

I am arguing here that the forces inherent in human actions embody and enact human habits, just as human character directs and informs human behavior. The Vatican is standing on the higher ground in this skirmish by insisting that humans need fidelity for the begetting of their children and the random fortunes of fertility to make their marriages generous. One appreciates the wisdom that led Roman teaching to allow for birth control, but one regrets the failure of nerve that led it to condemn contraception as an intrinsically denatured act rather than as a tactic of selfish habit.

A church which confesses that acts are good or evil insofar as they enhance or damage character must rely heavily on its cumulative experience of what happens to people who perform certain actions. The abundant testimony of credible believers has been suggesting that self-sacrificing Christians who have made themselves generously open to spouses and children, neighbors and strangers apparently moderate fertility contraceptively and yet thrive morally. This implies something about the act itself, about the full claims of sexual responsibility, and about the greatness of character it both requires and fosters. I would anticipate that consultation with the faithful will suggest as well that in the context of a self-sacrificing marriage, recourse to assisted fertility (AIH—artificial insemination—or IVF), if available without destruction of offspring, may also be a morally good act.

What the Vatican actually said about the matter is this:

> By comparison with the transmission of other forms of life in the universe, the transmission of human life has a special character of its own, which derives from the special nature of the human person. "The transmission of human life is entrusted by nature to a personal and conscious act and as such is subject to the all-holy laws of God: immutable and inviolable laws which must be recognized and observed. For this reason one cannot use means and follow methods which could be licit in the transmission of the life of plants and animals [*Mater et magistra*]."[34]

Much as most people admire Pope John XXIII, from whom that quotation was taken, we get nervous at talk about "immutable and inviolable laws." The agrochemical industry felt the same way about Rachel Carson's talk of the inexorable.

34. *Instruction*, introduction, 4, par. 3.

What the Vatican might have written could go something like this:

> The procreation of a person is unlike any other productive activity possible to human beings. Human production is an act of mastery. Whether it be in manufacture or in artistry or in agriculture or in animal husbandry, we produce objects, artifacts, and organisms that belong to us as the fruit of our handiwork. But with the yield of our bodies it is starkly different. All children are begotten, not made. We receive them; we dare not make them; we must not possess them. As our companions, children are given into our care and nurture but not into our dominion.
>
> Our very survival, as parents and as a race, depends on the generosity with which we make room in our world for every newcoming child, with deference and fellowship. All the advantage and power we enjoy because of our seniority must be restrained from exploiting the young. It must be turned instead to the work of guardianship.
>
> Now what is it about this recent program of technical assistance to people in their parenting that seems so perverse to us?
>
> It makes little difference whether a given individual will actually resort to any particular procedure, but across the world today men and women (proportionately to their affluence) are commonly invited to engage in a variety of acts:
> - to embrace one another sexually without leaving themselves open to children;
> - to conceive without any sexual union between the parents;
> - to examine children once conceived and, once born, decide whether to shelter them;
> - to destroy undesirable children by expulsion or dismemberment or poisoning, or by "putting them to sleep";
> - to purchase children from poor women when unable to bear them oneself;
> - to pay others to rear children when neither parent will agree to nurture and nourish them.
>
> What is wrong with that? The child has become a chattel, a commodity, a pet, a possession. Under the equivocal pretext of "producing only wanted children," men and women

have denatured their awesome faculty of procreation into one of manufacture and possession.

What we behold is an entire generation of adults poisoning its own capacity to parent. Marriage and the marriage embrace are no longer a single promise and a single acceptance of both spouse and children. People imagine that commitment to a beloved spouse can be made fast without enclosing within it a commitment to beloved children. It is our observation that they are doing violence to themselves and to their children.[35]

To the extent that biotechnology can truly facilitate the marital union—which is a single but twofold bond between spouse and spouse and between spouses and children—then we value it as art enhancing nature. But where it overrides our natural human makeup and relationships and fulfillment, it is an intrusion. Technology is proposed as the instrument of human self-determination, freedom, and choice, but if it is used to subvert our inherent needs and obligations, it will surely handicap and even destroy us.

But this only leads us to the gaping question that Rome has not yet resolved: How do we discern when technology is enhancing nature? How do we differentiate what is appropriately artificial from what is inappropriately artificial?

The Instruction tells us that artificial interventions into procreation and gestation must be morally evaluated insofar as they serve the dignity of the human person. According to Rome, this requires that procreation occur within marriage and that it follow from sexual intercourse that honors its two linked aspects: unitive and procreative.

Here one can look more closely at the document's treatment of artificial fertilization involving husband and wife (through either IVF or AIH). IVF is disallowed as it is presently performed because a number of eggs are extracted and fertilized, and then the excess eggs are discarded when conception succeeds. But even if and when the procedure is perfected so that such destruction of

35. See, on a similar note, Richard Stith, "Thinking about Ecology," *The Cresset*, Nov. 1981, pp. 7-10.

human beings is avoided, the Instruction stands by three moral requirements for rightful conception:

1. It must respect the inseparability of the unitive and procreative meanings of the sexual act. Contraception is rejected because it involves sex that is intended to be unitive but not procreative. IVF and AIH are rejected because they involve separate sexual acts that are procreative but not unitive.

2. It must be an act that is embodied as human beings are, with harmonious involvement of both body and soul. Fertilization outside the human body is therefore not acceptable.

3. Conception may be aided but never dominated by outside assistance. Neither IVF nor most AIH makes possible procreation in conformity with the dignity of the person. The child must be respected and recognized as equal in personal dignity to those who give him or her life. "The human person must be accepted in his [or her] parents' act of union and love; the generation of a child must therefore be the fruit of that mutual giving which is realized in the conjugal act wherein the spouses cooperate as servants and not as masters in the work of the Creator, who is love.

"In reality, the origin of a human person is the result of an act of giving. The one conceived must be the fruit of his [or her] parents' love. He [or she] cannot be desired or conceived as the product of an intervention of medical or biological techniques; that would be equivalent to reducing him [or her] to an object of scientific technology. No one may subject the coming of a child into the world to conditions of technical efficiency which are to be evaluated according to standards of control or dominion."[36]

The Instruction is arguing that technical assistance to a husband and wife in the course of their sexual union is damaging when it involves a generative act that does not involve the marital embrace. And because technical skill is so essential, that skill is said to dominate the couple or to assist them in dominating their offspring.

36. *Instruction*, II.4.c.2, 3; gender inclusion added.

Here, I suspect, some good principles may be getting a careless application. The generative act is being viewed as an isolated event, separate from the sequence of sexual union that the married couple has enacted all along. And we are not given a principle adequate to discern when technology is assisting and when it is intruding.

The Instruction argues repeatedly that the only context for moral evaluation is the individual sexual act, not the totality of a couple's conjugal life. Biomedical intervention which might be considered an assist to the long sequence of sexual union is accused of being an intrusion when judged within the single event. The very same point was argued in 1968 in *Humanae Vitae*, Pope Paul VI's encyclical on birth control. When that document appeared, I made the following observation:

> According to the ethical model followed by *Humanae Vitae*, one must assign moral value to methods of contraception within the isolated context of a single event of coitus, rather than the full sequence and story of love and childbearing throughout the course of a marriage. The Pope parts company with his advisory commission, which reported: "The morality of sexual acts between married people takes its meaning first of all and specifically from the ordering of their actions in a fruitful married life, that is, one which is practiced with responsible, generous and prudent parenthood. It does not then depend upon the direct fecundity of each and every particular act."[37]

Oliver O'Donovan, a professor at Oxford, has more recently addressed this subject:

> What upset so many in the teaching of Paul VI's encyclical *Humanae Vitae* was that it seemed not to perceive the difference of structure between sexual relations in marriage and simple fornication. Chastity in marriage was analyzed into a series of particular acts of sexual union, a proceeding which carried with it an unwitting but unmistakable hint of the pornographic. A married couple do not know each other in isolated moments or one-night stands. Their moments of sexual union are points of focus for a physical

37. Burtchaell, " 'Human Life' and Human Love," *Commonweal*, 15 Nov. 1968, pp. 248-49.

relationship which must properly be predicated of the whole extent of their life together. Thus the virtue of chastity as openness to procreation cannot be accounted for in terms of a repeated sequence of chaste acts each of which is open to procreation. The chastity of a couple is more than the chastity of their acts, though it is not irrespective of it either.[38]

The very same issue is raised by the Vatican rejection of IVF and AIH. And on this matter O'Donovan has made a like observation:

> There are *distinct* acts of choice, which may involve persons other than the couple, in any form of aided conception, including those forms of which [Catholic official opinion] approves. Whether they are *independent* acts of choice is precisely the question which requires moral insight. If they are indeed independent (and not subordinate to the couple's quest for fruitfulness in their sexual embrace) then they are certainly offensive. But that point cannot be settled simply by asserting they are distinct. The question remains: is there a moral unity which holds together what happens at the hospital with what happens at home in bed? Can these procedures be understood appropriately as the couple's search for help within their sexual union (the total life-union of their bodies, that is, not a single sexual act)? And I have to confess that I do not see why not.[39]

There is, I would say, good reason to consider contraception, IVF, and AIH as capable of enhancing the natural course of a marital life in the same way that a cesarean section or carefully disciplined ("natural") birth or bottle-feeding with special supplements do. There can be artifice and technology that enhance nature. But that needs to be evaluated within the full continuity and integrity of a couple's sexual life. The moral worth of technical intervention would derive from whether or not the union itself was generous between the spouses and toward offspring. It is, after all, a union with many encounters. The Instruction points out early in its argument that since humans are persons, our sex is quite dif-

---

38. O'Donovan, *Resurrection and Moral Order* (Grand Rapids: Eerdmans, 1986), p. 210.

39. O'Donovan, *Begotten or Made? Human Procreation and Medical Technique* (New York: Oxford University Press, 1984), pp. 77-78.

ferent from that of other animals, and its biological aspects must be viewed in the light of its personal aspects.[40] And the fact that individual acts of sexual intercourse are, for a human couple, incidents within a continuing story obliges us to assess biomedical interventions insofar as they serve the saga, not just the episode. Steadfast attention to the whole . . .

No one will accuse Rome of not having been steadfast. Indeed, in the longer run of history it may emerge that Rome's disdain for most modern efforts to govern the procreative power of sex was the earliest campaign of resistance to the dogma that human choice need encounter no limits to its dominion over nature. More is the pity that this resistance movement has been a stammering effort, blessed with little of the convincing rhetoric of its younger sister, the ecological movement.

Journalists reporting immediate reactions of Catholic couples to the Vatican Instruction sympathetically presented those involved in the new reproductive technologies as outraged that their well-intentioned desire for children should be called into question:

> Two Catholic couples turned on their TVs last Tuesday night and learned that they were sinners. . . .
> "It [IVF] was just a little helping hand," Doug says of the procedure.
> "Pronouncing this a sin is almost unbelievable," says Michael, 31, who was educated in parochial schools. "When I think of sinning, I think of murder and lying and cheating—not of all the happiness I get from Lindsay. I just don't think it's morally wrong. I guess you have to have your own standards about this."[41]

> John Pieklo, 40, of Perrine, and his wife, who is 35, just had a healthy baby in December. Although they are Catholic, his wife had amniocentesis with the idea that she would have had an abortion if there had been a severe problem with the baby.
> "I do realize this is against our church principles, but I also feel it is very important that we raise a healthy child," said Pieklo. "If it had a brain problem or handicap problem

40. *Instruction*, introduction, 3, par. 2.
41. Sandra Rubin Tessler, "Little Bundles from Heaven Are Theological Nightmares," *Detroit News*, 15 Mar. 1987.

before birth and could have been aborted, I'm sure we would have aborted. . . . Raising a child is hard enough without putting yourself in the position of raising a handicapped or brain-damaged child."[42]

In reading these remarks I was reminded of correspondence I had had with a civil engineer about the depletion of the Ogallala Aquifer. I had oversimplified the groundwater crisis in the Southwest, he told me:

These farmers are not, as a whole, greedily consuming water with total disregard for the present and future good of the world. . . . Through enhanced understanding of the limitations of man, technology and nature, the farmers of the desert Southwest hope to be able to provide grain, cotton and cattle to an ever increasing world population.

To say that the role of the farmer in drawing water from the Ogallala is a sin is, I believe, a mistake. Many of these farmers are deeply religious people.

This is a pervasive modern moral argument: that the moral worth of our actions derives not at all from their inherent direction and force but is infused into them by our intentions. The tradition, of course, has doggedly pointed out that there is as well a reverse movement: our actions have an inherent power to transform our attitudes by enhancing them or by enfeebling them.

Perceptive critics within and without the Catholic Church, though dissenting from what they consider an indiscriminate application of principles by the Instruction, commend the Vatican for insisting that much more is at stake than a good-natured wish to have one's own, healthy child. Columnist Charles Krauthammer sees Rome as resistant to three scientific innovations: synthetic families (surrogate parenting), synthetic children (test-tube babies), and synthetic sex (artificial insemination). In general, they pose a threat to the integrity of the family, the dignity of the individual, and the personal bonds of sex.[43] Sidney Callahan writes,

42. Cathy Lynn Grossman, "Many Catholics Dispute Vatican Edict," *Miami Herald*, 13 Mar. 1987.

43. Krauthammer, "The Ethics of Human Manufacture," *New Republic*, 4 May 1987, pp. 17-21. Krauthammer had observed earlier that the decision of the trial court to uphold the Whitehead-Stern surrogacy

"In the case of reproductive technology, ethical positions best emerge from the question: What will further the good of the potential child, the family, and the general social conditions of child-rearing, while strengthening our moral norms of individual responsibility in reproduction? We should move beyond a narrow focus upon either biology or individual desires for children."[44]

In the *Instruction on Respect for Human Life*, Rome is addressing the issue of procreation on broader terms than most other parties to the conversation. Yet something is still missing. The See of Peter, which ought to be drawing on the wisdom from its network of compassionate and insightful pastoral interchange across the world and through the centuries, is not speaking out of the only reliable source of insight it possesses: an inspired, imaginative, and attentive reflection on experience. Instead, Vatican theologians may have had their eyes too close to their task, and their task has been on paper. And that, ironically, is a failing endemic to technology.

The Roman Instruction has also been berated for having urged that civil laws protect both individuals and families from the clumsy hands of irresponsible technology. Much outcry about priestcraft and coercion has been heard.[45] Yet the very moral questions raised by the Vatican have been echoed by others whose call for legislative regulation has never been criticized. Listen to one psychiatrist writing about the famous Baby M case in the *New York Times:*

> When the child learns, as she inevitably will, that she was conceived to be sold, what will she feel and think about her-

---

contract and to award Baby M to her father and his wife had ironically honored two of the very principles cherished by Mary Beth Whitehead's feminist supporters. The decision to consider the surrogacy contract valid was harmonious with the doctrine that women should have control of their own bodies. "If they should have the right to terminate the life of a fetus, how can they be denied the right to grow one for a fee?" The ruling that the child's father was the fitter parent (against the older presumption that the maternal bond made the mother the more nurturing parent) was at peace with the feminist claim that biology is not destiny, and fathers can and should be nurturers of children as much as mothers. See "Feminists Helped Make Rules, 'Lost' by Them," *Chicago Sun-Times*, 6 Apr. 1987, p. 25.

44. Callahan, "Lovemaking and Babymaking," *Commonweal*, 14 Apr. 1987, pp. 233-39.

45. See Maguire, "Vatican Birth Decree Casts a Long Shadow."

self and her natural mother, Mary Beth Whitehead? The child will understand that the bottom line—which cannot be erased or rewritten—is that the natural mother, her own mother, sold her. What is the fair price of this merchandise? ...

Mrs. Whitehead's children saw their mother carry and deliver a child like themselves—a half sister—whom the mother then gave or sold to another family. All children fear that if they are bad they will be sent away or abandoned. Won't these children fear that their mother might sell them, too? ...

Another issue, obviously undemocratic in a democratic society, is the exploitation by a middle-class well-to-do couple of a woman of a lower socio-economic class. The implication clearly is that the richer, better educated couple— Dr. Elizabeth Stern and her husband William—can offer more (at least materially) and that the child's "best interests" thus would be served by awarding the child to that couple. This slippery slope can lead only to racist and supremacist rewriting of laws and controlling of human rights. ...

But what of a man's desire to have a child who will carry his own genes? While the feeling is understandable, it has elements of narcissism and ego that have more to do with the man's needs than the child's best interests. ...

Society is faced with a situation in which biotechnology and commerce conspire to invent a complicated transaction that leaves little room for ethical or moral values. The fact that something can be done does not mean that it must be done. "Surrogate mothering" should be outlawed. ...

Morality, ethics and values are not a religious matter alone. They determine how we live and how far we go in reaching our full potential as humans.[46]

This is plain talk by a citizen arguing that an immoral form of widespread personal abuse ought to be curbed by the law of the state. Surely the same suggestion is not any less allowable if the leaders of a church are making it.

Rachel Carson defended her work in these words:

46. Robert E. Gould, "And What about Baby M's Ruined Life?" *New York Times*, 26 Mar. 1987, p. Y27.

Now to these people [defenders of clinical manipulation of nature], apparently the balance of nature was something that was repealed as soon as man came on the scene. Well, you might just as well assume that you could repeal the law of gravity. The balance of nature is built of a series of inter-relationships between living things, and between living things and their environment. You can't just step in with some brute force and change one thing without changing a good many others. Now this doesn't mean, of course, that we must never interfere, that we must not attempt to tilt that balance of nature in our favor. But when we do make this attempt, we must know what we're doing, we must know the consequences.[47]

One cannot put words into Cardinal Ratzinger's mouth, even elegant words like Carson's. But one hopes that this is what he also would have to say.

The curial Instruction, like the encyclical *Humanae Vitae* of two decades earlier, seems to stumble by locating the pivot of moral meaning in the invariable linkage of procreativity with each in-dividual act of intercourse. Neither document allows the single sexual act to be significantly coupled into the long train of sexual union between spouses. That has been and will continue to be a vul-nerable point at issue among Christians who share the Vatican's overall moral imperatives. It is those prophetic imperatives, insist-ing that the personal dignity and bonds and loyalties which differ-entiate human mating and begetting from those of unspirited animals, which will survive their fumbled applications here.

One is sometimes led to think that such divine guidance as Rome enjoys comes more in the form of stubborn intuition than as articulate insight. But in this matter of the love between hus-band and wife, and the love between parents and children, the natural needs of men and women and children have to be bound together through sexual intercourse . . . though it sees through a glass darkly, Rome does attend steadfastly to the whole.[48]

47. Carson, cited by Carol B. Gartner in *Rachel Carson* (New York: Frederick Ungar, 1983), p. 107.

48. An earlier version of this essay appeared in *The National Catholic Reporter* and appears in this form by permission of its editors.

The Story of an Encounter

Richard G. Hutcheson, Jr.

REFLECTIONS ON TECHNICAL CHANGE AND MORAL JUDGMENT

More than twenty scholars—representing fields as diverse as philosophy, theology, psychology, sociology, political science, medicine, and law—gathered in New York City in late April 1988 for the first of two invitational conferences, both entitled "Guaranteeing the Good Life." The moderator of both conferences was Richard John Neuhaus. Behind the conference discussions lay the return to prominence, in the late twentieth century, of eugenics—the movement to attempt to improve and even perfect the human species by technological means. An article by Neuhaus called "The Return of Eugenics," which was published in the April 1988 issue of *Commentary* and which appears in this volume, provided the background for the conference, although the article was not explicitly discussed in this context.

In his introductory remarks, Neuhaus dedicated the conference to Paul Ramsey of Princeton University, who had participated in its early planning prior to his death in February 1988. "There are few people—one would almost say there is nobody else—in the last two decades of American intellectual life who have made as powerful a contribution as Paul Ramsey to getting us to reflect on where we're going with regard to technology and cultural and moral assumptions," Neuhaus said.

A little later Neuhaus spelled out the agenda of the conference: "I envision this conference as responding to what I expect most of us perceive as a truly threatening convergence of practices, both existing and proposed, having to do with the assertion of human control, mainly through technology, over the contingencies of the human condition. This convergence, I think, is accurately described as 'the return of eugenics,' with all the freight that bears in terms of the Nazi experience. I think it is not at all alarmist but truly alarming to see the ways in which the language, the logic, the stated purposes, and so on are so disturbingly analogous. So the question for our time together, it seems to me, is whether the progress of this mounting threat is as inexorable as it seems at times to be. It is not apparent to me that we are culturally, morally, or spiritually equipped to have any confidence that there are some things that we, at the end of the twentieth century here in the United States of America, just won't do. So to find out whether or not that rather doleful suspicion is justified is, in large part, what I would like to see come out of this conference on guaranteeing the 'good life.'"

Stanley Harakas of Holy Cross School of Theology (Greek Orthodox) raised a question about the title of the conference, "Guaranteeing the Good Life." " 'Guaranteeing,' I assume, directs itself to technological kinds of guarantees," he said. "In other words, it involves the ability of science to manufacture something. The 'something' is what concerns me in terms of an ethical assessment of 'the good life.' 'The good life,' to me, is essentially a teleological statement. You are positing something and characterizing it as good, as an end that is to be attained, that this guarantee is designed to produce. In a society like ours what is probably meant is a kind of existentialist, individualist, personal choice. If that choice can be assisted by technology, it is 'the good life.' And the question then becomes, 'Is that, in fact, the nature of the good life?' I myself cannot subscribe to that, and I think that in part the existential presupposition that some of you have commented on in your papers is precisely the fundamental concern. So what I want to do is raise the question of what we mean by that 'good life' and how it is produced."

"The title is designed to be exquisitely ambiguous," Neuhaus responded. "The 'guaranteeing' part refers not simply to technology but to human control over the contingencies that get in the way of what people think is the good life. But the assumption with respect to 'good life' is that in a society such as ours there are numerous definitions of what constitutes the good life that are consciously or unconsciously in play. And before we're finished here we should indeed get some fix on the culturally dominant notions of the good life, and whether those ideas do not play into the drive of the first part of the title, to 'guarantee' or to control whatever contingencies get in our way. All those ambiguities are in the conference title by design."

Categorical Principles as the Basis for Moral Judgments

The first part of the conference focused to a considerable extent on the philosophical presuppositions underlying moral judgments. Hadley Arkes of Amherst College, whose paper had previously been read by all the conference participants, made some opening remarks in which he emphasized, in Kantian terms, the universality of moral principles on the basis of which judgments are to be made. Technical changes, he said, do not alter the domain of moral judgment, which is based on categorical principles. The question of the taking of human life, for instance, is not affected by the technology of the means used. Recognized universal principles, Arkes went on to suggest, become incorporated into the law, and thus become enforceable.

Against this background, Neuhaus sought to focus discussion of the Arkes paper directly on the modern technological interventions in human life that raise the question of moral principle. Abortion, he suggested, is the most fevered issue. Assuming a common agreement among conference participants that the abortion debate engages a moral dimension and that the abortion issue is not an isolated question, he asked, "What are we to make of the fact that these issues are commonly posed as questions raised by technological advances?"

As the group began discussion, however, the theoretical basis for moral judgments, not the triggering issues, became the first focal point. Jean Bethke Elshtain, a political scientist from the University of Massachusetts (and now at Vanderbilt University), questioned the Kantian framework of the Arkes paper. "What about exceptions to principle?" she asked, citing as an illustration the just-war theory, in which killing is justified under certain circumstances. Further, she raised a question about the circumstances under which principles should be enforced by law, citing Prohibition as an illustration of enforcement of moral principle by law with deplorable results. The *Roe v. Wade* Supreme Court decision, currently having the force of law regarding abortion, may be overturned by a future decision, she noted.

Richard Stith of the School of Law of Valparaiso University spoke next, suggesting a possible compromise position, a middle ground, on the enforcement of moral principle by law. "There are some things that may not be justified but can be excused," he said, pointing to the European tradition of constitutional principles that do not necessarily call for punishment when violated. Such a position is more nuanced than is customary in our society. With such an approach, he explained, abortion might be regarded as illegal (establishing the moral principle) without an accompanying enforcement.

Arkes, responding to Elshtain and Stith, agreed that there might be exceptions to principles. "It is plausible to ask whether or not there may be compelling grounds for the taking of life," he said. But he maintained that the principle against the taking of innocent human life is still intact even if a middle ground allowing for exceptions or non-enforcement is established. With a basic principle that "human offspring are not other than human," one may then ask, "What are the grounds for taking life?" In response to a follow-up question from Sidney Callahan of Mercy College, Arkes went on to reaffirm his basic position: "Principles are absolute, but they must be applied with understanding of the circumstances, based on experience."

A Contractarian Basis for Moral Judgments

Stephen Post of Marymount College (and now of Case Western Reserve University) raised questions about the meaning of "categorical" or "inherent" as applied to moral principles, then proposed a contractarian basis for moral judgments, under which "moral principles are agreed on by a particular society because they are necessary for peaceable, coherent existence." With some variations in terminology, this contractarian or communitarian view became the major alternative to Arkes's categorical prin-

ciples in the debate over the basis of moral judgments to which
much of the morning session was devoted.

Neuhaus, seeking once again to focus attention directly on
the morality of technological interventions in human life, suggested that how we arrive at moral principles may not be the most
critical question before this group. The question, rather, is how
we respond to particular challenges on the basis of those principles. Noel Reynolds of Brigham Young University dissented.
"How we arrive at moral principles does make a difference," he
told Neuhaus. "It determines how we make laws."

At this point Hadley Arkes re-entered the discussion, arguing against a contractarian view of the moral framework. There
are categorical assumptions about creatures with competence to
enter into contracts, he said. These provide the moral framework
within which contracts can be made. "There are certain kinds of
terms that a moral being cannot accept as part of a contract. The
law will preclude certain kinds of contracts as not fitting or
proper. For example, contracts for prostitution will not be permitted. Freedom of contract works within that domain of permissible, legitimate choice. The law must mark off the limits of what
is legitimate or illegitimate—in other words, mark off a moral
framework in which contracts can be made. But it is not through
contracts that we arrive at the understanding of what it is that establishes the sense of right and wrong which marks the limits. As
Daniel Webster and others have pointed out, this government did
not come into being through a compact—a contract. We ordained
an authority. We did not accede to a contract, because the corollary of 'accede' is 'secede,' to undo the contract. Contracts proceed
within a moral framework which is not itself part of the contract."

Reynolds assured Arkes that he was in substantive agreement with his position. Nevertheless, Reynolds proceeded to
challenge Arkes directly. "I think your argument is too easy; essentially it takes a cheap way out. We have a more difficult
problem than the one you described. And the problem is this business of 'categorical'—tieing morality and law into categorical
principles the way you've done. Moral principles enter the law
only through the mediation of human beliefs. Different people involved in the formulation of law will have different moral beliefs.
We can't presume all those moral views are finally correct. Certainly they have views that coincide to different degrees: the

views of voters, legislators, judges, everybody, at all stages of the political process. But these views are not uniform, not universal. When Aquinas formulated the purpose of law, the one you cited for us, that the law will lead us on to some kind of moral improvement, he couldn't have been thinking of the U.S. Congress, the Massachusetts legislature, or the Chicago city council. Our experience seems to be that law lags well behind morality, not the other way around, and that law is unavoidably a product of agreement among human beings. It therefore cannot be the case that if something is immoral by some correct standard of morality, it has to be illegal. That is too easy a connection."

"With reference to your comment about the Chicago city council," Arkes responded, "when you recognize a sick horse, it's only because you have an awareness already of what a healthy one looks like. You recognize a city council that seems to be defective because you have an idea of what a good legislature would be."

"Are you asserting that we all have the same base of moral discernment by which to judge?" Reynolds asked.

Said Arkes, "I'm saying that in a curious way you're presupposing the very argument you're resisting. In the political arena we assume that the polity has an authoritative agency which can make decisions that are binding on the population. We're not saying, 'Here is a code of conduct we think would be good to adopt, so everyone who agrees with us, sign on to this.' When we enter the polity and make laws, we say this or that policy will be imposed even on people who find it distasteful. I assume that is the mark which distinguishes the polity from other associations or authorities. This association may make laws that are universal. Some people are in a position to make authoritative decisions binding on others that displace personal choice. And we have to ask in a serious way, 'How do you justify that state of affairs?' When we begin to legislate about discrimination, do we have contingent goods in mind? Do we do it because it affects the reading scores of black students and their prospects for jobs? Is that good or bad for retail business? Does that lead to some other contingent good? We could go on up the scale of contingent goods to an infinite regress. Now whether you use Kant or not—let's take the label out—you seem to acknowledge that there are instances in which we would be justified in imposing a common policy on people and displacing their personal preferences. I don't care whether you happen to refer to such policies as 'moral' or say that they contain the attributes of a logical and moral proposition, but recognize that when you do that you probably explain the properties that attach to those propositions which you are willing to treat as enforceable as opposed to those you are not."

Arkes concluded by suggesting that it doesn't matter whether you attribute such reasoning to Kant or to someone named "Doodles Weaver," a remark which met with laughter.

At this juncture Neuhaus sought to clarify: "In response to the kind of question Noel raised, you're saying that the very nature of polity requires a principle of legitimation, and that whether or not you call it a moral proposition, it has to be something like a moral proposition."

Pursuing the issue further, Reynolds returned to an alternative view proposed earlier. "If you were to take the contractarian view that Stephen Post advanced—and I assume that isn't your view—that would provide an explanation of the basis of moral judgments."

Arkes, however, remained unconvinced. "Contracts *presuppose* a moral framework in which to justify the enforcement of contracts," he said.

Reynolds summarized the disagreement. "It's really just a difference between two different views of law, the one being a conventional view which you are anxious to avoid: the idea that the law is a matter of agreement within a society. The other is a view that authority and law are justified finally because they are morally correct, and that enforcement is justified finally because it is morally correct."

Neuhaus pressed Reynolds on this point. "Noel, even if we accept a purely contractarian approach—that law is what the society agrees on through certain contractual procedures—the agents in that procedure are clearly acting on their beliefs—which in ordinary language we would call moral beliefs—about what is right, what is wrong, what is useful, and so forth. So you are not disagreeing with Hadley's point, are you, that in the very nature of the law is the assumption of a moral framework? What you're accenting is that all the players relevant to the formation of the law may not agree on the definition of the framework."

But Reynolds persisted. "Although there may be great

similarities in the way some individuals see things morally, there are also great differences. When we talk about pluralism, that's what we're talking about. And it seems to me that Dr. Arkes's argument necessarily keeps sliding into a holistic notion, that somehow there is either holistic agreement or, in lieu of that, truth, moral truth—which is something more authoritative—in the legal realm."

Arkes sought common ground with an analogy of an anthropologist who presses for cultural relativism but concedes that although the notions of what is right and wrong may vary from society to society, all societies have the concept of right and wrong. "They have the same properties in that what is considered to be right takes precedence over what is wrong. Here we may find people thinking they seriously disagree about the foundations of our public policy, and yet the most decisive thing is what they are already taking for granted about the existence of the polity."

"And that there is something to disagree about—namely, right and wrong," Neuhaus interjected.

"But who defines what is right and wrong?" asked Brigitte Berger of Wellesley College (and now of Boston University).

"We can have one law, but we can have many views of what is right and wrong," said Reynolds.

"Sure," said Neuhaus, concluding the interchange. "And what becomes the law, at least in certain polities, are those views which prevail at a given time, through the procedures which people are not necessarily agreed on, but which they have no choice but to go along with."

The Kantian Framework

David Novak, now at the University of Virginia, challenged Arkes on his use of Kantian language while indicating a willingness not to be bound by the Kantian tradition. "The fact is that your argument is Kantian through and through, and has a number of the problems of the Kantian argument. Universalizability has to have a certain locus; there are certain things we want to universalize, and certain things we don't want to universalize. In Jewish tradition, dietary laws are a categorical imperative; everybody who is under the covenant of Israel absolutely must not eat pork. That would be trivial in other contexts, but in the Jewish context it is a categorical imperative. Universalizability alone, as W. D. Ross has indicated, is a very weak point of Kant. But the universalizability that you want, and that Kant wanted, is determined by the unchanging, immutable nature of both the moral subject and the moral object, who turn out to be the same—namely, the human person. For Kant, the human person is a moral subject and ergo a moral object because the human person is *causus sui*, an end in itself, with qualities of both rationality and will.

"Based on the criteria of universalizability, there are a number of things that you could simply eliminate as immoral. For example, on those grounds you could eliminate genocide. But your prime example of abortion simply doesn't work if the definition of the human person is of a rational, human *causus sui*, a person capable of causing and being responsible for his or her own acts. Granted, I could indicate that a kind of gratuitous abortion might be offensive to the moral sense. But if a mother says, for instance, that her pursuit of a Ph.D. in philosophy depends upon aborting her fetus—and let's say she has to make a choice between those two perceived alternatives—then clearly on the grounds of her being a rational person she could justify, by these criteria, aborting her fetus so that she could continue her studies and not end up changing diapers and breast-feeding."

"David, are you saying that such conclusions are implicit in Hadley's assumptions?" Neuhaus asked.

"I like Hadley's practical conclusion," Novak answered, "but I have problems with the philosophical undergirding of that conclusion. That's the point I'm making. The Kantian metaphysic of morals is insufficient to produce an absolute ban of such things as abortion, infanticide, and even euthanasia. In cases in which the subject is comatose or pre-rational, then very clearly I don't think the categorical imperative obtains."

"Why can't I say that it's wrong to take human life without justification, and that that's a universal moral proposition, and that I make the proposition because God has revealed it to be his will?" Neuhaus asked.

"I would agree with that wholeheartedly," said Novak. "But you and I are not Kantian. Hadley is. According to a Kantian definition of human nature, a fetus is not human. A fetus does not possess intellect and will."

Neuhaus then asked Arkes, "Does your understanding of universalizability entail the anthropology that David just described?"

"No," he answered. "If we strip away the labels, and we say 'What do you understand to be the nature of this enterprise we are engaged in?' do we find ourselves, as Kant says, seeing philosophers offering reasons at their highest level in order to prove to themselves that there are no reasons? We find people engaged in offering reasons to one another about matters of right and wrong, what is desirable and what is undesirable. When they do that, do they think they're speaking merely about matters of tea and coffee and personal taste, or do they think they're speaking about things that people ought or ought not to do, where oughtness attaches, where restraints may be applied, where law may be invoked? If they do that—without invoking Kant or anyone else—do they understand that there are certain things which are good only as means toward some other end, or certain things which are wrong in themselves? Is the wrong of genocide that it is bad for retail trade? Or is there something inherently wrong with killing masses of people? Whatever label you finally attach to these things, and regardless of whether you think I'm making an illicit move from Kant (which I would certainly dispute), I think you yourself would be faced with the same questions and problems once all the Kantian trappings were removed."

Neuhaus summarized the point: "So regardless of what philosophical school is invoked, in a polity people are engaging in these kinds of discussions and making those kinds of decisions."

Without a clear resolution of the question of universal principles versus contractarian decisions as the basis for moral judgments, much of the remainder of the morning's discussion focused on specific moral issues. Father Francis Canavan of Fordham University, although he supported Arkes's claim that there are universal moral principles, turned the discussion toward society's moral consensus, however based. Law, he said, must be based on public moral consensus. "The undeniable fact is that human beings are incurable moralizers," he said. "We think morally—even Stalin felt obliged to give moral justifications for the kinds of things he did. Most people do make moral judgments, and we do relate them to moral laws."

Referring to Jean Bethke Elshtain's earlier remark on efforts to overturn *Roe v. Wade*, Canavan said, "If the day were to come when the Supreme Court would simply say that *Roe v. Wade* was wrongly decided, that the 14th Amendment does not prohibit states from regulating or prohibiting abortion, what would happen? Well, that in itself wouldn't prohibit abortion at all. It would simply return the question to the political process, where I think it really belongs. It would return the question to fifty different states, so you would get different answers. This is quite distinct from a constitutional amendment prohibiting abortion, with or without some exceptions.

"The reason the amendment prohibiting alcohol failed was that there wasn't a sufficient national consensus behind it. We have prohibition in this country—of narcotic drugs. The prohibition results in widespread corruption, organized crime, and all that, but we aren't prepared to revoke the prohibition merely for that reason. Maybe we should; some people are saying so. But I don't think that saying 'Prohibition didn't work' settles anything.

"Still, a reason for not prohibiting abortion might be that it simply wouldn't work. The law might break down because of the absence of a sufficient supportive consensus. Now I'm not saying that democratic majorities make moral law. They don't. What I'm saying is that the legislator has to take into account a number of factors, one of which is whether the law enjoys sufficient support to be enforceable. That doesn't say we should maintain *Roe v. Wade*. It says that sufficient support would be one of the things that would have to be thought about when considering the constitutional prohibition of abortion. That would be an ideal that I think would have to be worked toward through a long process of changing people's minds.

"The other phrase I want to comment on is one used by Richard Stith. He said you might want to say that abortion is against the law but gets excused. On the face of it, this sounds somewhat silly, but in fact it is not. Two years ago the Supreme Court upheld a Georgia law against sodomy. Now no one was punished; the law wasn't enforced. So what difference did it make?

"I can give you an illustration. The effort to get the courts to recognize the right of homosexuals to adopt children succeeded in a number of states. One state in which it didn't succeed was

Virginia, where the Supreme Court said, 'Sodomy is a crime in Virginia, and therefore it cannot found a legal right.' This figures into the abortion issue: if a thing is illegal, it isn't a sufficient answer to say that the law really isn't enforced and doesn't carry penalties. The law may be blocking out all sorts of additional rights. We see this with regard to the abortion question and other moral questions."

The Complexity of Technological Issues

Bernard Nathanson, a physician, brought into the discussion the attempts by technocrats to set moral framework on technological grounds. He produced a recent publication entitled *Ethical Considerations of the New Reproductive Technologies*, put out by the Ethics Committee of the American Fertility Society in response to the *Instruction on Respect for Human Life in Its Origin and on the Dignity of Procreation* issued by the Vatican's Congregation for the Doctrine of the Faith.

"It is the technocrats' wonderfully lucid and seductive response to some of the questions you have raised, Professor Arkes," said Nathanson. "It sets aside a morally neutral zone for all of this new reproductive technology, including early abortion. It defines a new state of human existence called the 'pre-embryo.' This is the embryo up to two weeks after fertilization. And it indicates that in that morally neutral zone anything goes. This is not a human being; it is merely a 'pre-embryo.' In my Williams obstetrical textbook—the seventeenth edition, which just came out last year—I looked up the definition of a 'pre-embryo,' and it doesn't appear in that book, or in others. Yet these pronouncements by technocrats, it seems to me, are in effect more authoritative than pronouncements on these subjects by politicians."

Commenting further on the complexities of technological issues, Nathanson went on to say, "Some years ago I stated that the definition of abortion is not killing; the medical definition is purely the separation of a product of conception from the mother prematurely, before twenty weeks. The use of cryogenics—flash-freezing the product of conception, removing it, and transferring it to another woman or some sort of life-support system—is becoming an increasingly workable technology (it is being done experimentally in veterinary medicine). A certain number of these embryos will die in this process. Is this abortion? Should someone who engages in this process be regarded as guilty of some sort of homicide? With in vitro fertilization (IVF), a certain number of four-cell embryos created in the test tube die in the process of being either cultured or transferred. Is this in fact abortion? Should we prosecute on these grounds if we are legislating against abortion? There is a proven method of sex selection that filters out sperm of a certain type—carrying, let us say, the Y chromosome—and uses them for fertilization. The method is virtually guaranteed. The question is, if it fails, as it does in about 10 percent of the cases, is the donor entitled to abort that child?"

Neuhaus sought clarification: "You're indicating a number of these technological developments. Is it your own intuition or conclusion that in fact they create situations and contingencies that affect what you believe to be applicable moral propositions?"

"Yes, I do, absolutely," Nathanson replied. "I think the new reproductive technologies are taking us into gray areas and unexplored areas that challenge us enormously in terms of moral judgments."

Moral Judgments regarding Abortion

Fred Sommers of the philosophy department of Brandeis University continued the discussion by suggesting a different basis for moral judgments on abortion. "Earlier there was a question about what attributes make a person. This question is common to those who maintain that abortion is wrong because it is killing, homicide. But there are other positions. Would a separation of the fetus be wrong if it did not involve killing? Separation might be wrong for other reasons. Here I want to talk about philosophers who don't agree that what exists from the moment of conception is a human being. At this juncture technology comes in to attempt to help decide when so-called person-making characteristics exist beyond the stage of conception. According to some philosophers, we are beginning to know enough about neurology so that we could reach a point when we would know that what exists is a person. Others maintain that viability itself is a shifting thing because new technology establishes viability earlier and earlier. And still others maintain that personhood is only arrived at about a year after birth, so that infanticide as well as abortion would be permitted. "I want to point out that if this is the question, there are all these possible answers, both theological answers and various technological answers, that distinguish between existing as a human being and existing as a person?

"But even if abortion isn't considered a form of homicide, it might still be considered wrong because there's the social context. If conception has taken place, the parents are responsible for the life they have created. One of the strategies of those who favor abortion is to dismiss this, saying that the mother has no special responsibility. I'm thinking of that famous article by Judith Thompson, in which she says that the mother has no special responsibility unless she's decided to have the baby. Her only question, then, is 'Is this a person?' and she says that even if it is a person, there may be other overriding considerations—a woman's rights over her own body and things like that. What's overlooked by people like this—and, I think, even by theologians—is the special responsibility the parent has. This responsibility could be invoked even in the case of separation, even if it's not a case of killing. We might say, 'You have no right to do this because you've undertaken something and you have to carry it through.' I want to bring in this social or cultural aspect, this institutional aspect, of the matter, because it rescues the abortion debate from being at the mercy of this or that criterion about when human life or personhood begins. If you're not a theologian and you don't believe personhood starts with conception and you still want to have a concrete position on abortion, you may have it on the basis of this social-institutional consideration."

"The framework of institutions and other relationships?" Neuhaus asked.

"Yes. Having a family, being part of a family, becoming pregnant are crucial to our thinking here. The presumption is that the mother is responsible—not only for the health and survival of the child after it's born but for the health and survival of the family. Obviously the family as social institution is incompatible with abortion. The family is a social situation—a very intimate social situation. This gives the mother responsibility for the survival of the child."

"So the critical question," Neuhaus inquired, "is not what is in the mother's womb but what is entailed in being a mother, and the obligations that attend that identity?"

"That's right," Sommers agreed.

At this point Brigitte Berger returned attention to Arkes's paper. "In this paper, which I think is of great interest, there are basically two agendas. The first agenda is to give to the fetus the status of a human being. I must say that, for me, is a problem. I'm glad Richard Neuhaus hasn't forced us to take a loyalty oath on certain positions, so I didn't have to come out with where I stand on the abortion issue. I have a lot of problems with the abortion issue. I don't think the question is as simple as it has been made out to be, because in the end the question really is, When does human life begin, what is the genesis of personhood? I'm not as certain as Hadley Arkes seems to be that personhood begins at the moment of conception. But that is not the issue at this moment. Establishing that the fetus is a human being is the first agenda he has.

"The second agenda is that of the consummate jurist: to try to find vehicles to influence Supreme Court decisions on the revocation of *Roe v. Wade*. And there the question is procedure, and on what basis one would make one's moral argument. In a situation where there is a lack of moral consensus, the question becomes, in the end, Who has the power to define what's what? Who does what to whom? But the real contribution here, it seems to me, comes from the technologists. And here I listened with great interest to Dr. Nathanson, because the argument among technologists of medicine is very different. They don't have a moral argument at all. Let the medical man do his business, they say. Here Hadley has made a great contribution. One thinks of Eichmann in Jerusalem, when he said, 'I didn't kill anyone at all. I was just a superior railroad connector, making connections between different trains.' But that doesn't absolve the technician from moral responsibility."

"So the potent point of Hadley's that you're affirming," said Neuhaus, "is precisely that those are moral questions."

"Yes, but let me get to the real point," Berger continued. "Even though I have real problems with the definition of human life that Hadley proposes, I also think that *Roe v. Wade* is wrong. I wouldn't rely that much on the state's definition of the fetal process, or the scientists'. In the end, morality is a question of religion. But if law is the normative standard, then the law excuses everything. For the doctors it's convenient to have a law; then they can do their stuff. It absolves people from their own responsibility. In the end I think abortion would continue even if the law was changed, but it would no longer absolve people from their responsibility. What we have done in modern society is to let people hide behind the law."

"So were there a law, at least people would know they were making a decision of moral consequence," Neuhaus concluded.

Bernard Nathanson re-entered the discussion, picking up on Berger's reference to technologists. "The morally relativistic position of reproductive technologists is that if one presupposes that there is a right to abortion, to destroy the life, then there is an equally inviolable right to procreate life, irrespective of the consequences."

"That's the reason I said that in the end it's a question of the way we define personhood," said Berger.

At this point Arkes responded to Berger. "I don't think I said I was arguing that the fetus is a person. I don't know whether it's a person. The argument over personhood is simply an argument over whether this being—whatever you want to call it—has a claim to protections of the law. Or does the law give us a reason for saying that there may be frivolous reasons for the taking of that life? Years ago my friend Daniel Robinson made an interesting argument on the basis of 'animal liberation.' He maintained that it was possible to protect the fetus simply by seeing it as an animal, without even getting into the question of whether it's a human being. The same propositions that protect animals from the frivolous taking of their lives would also protect the human fetus. When we raise these questions, asking what there is about this being that makes it deserve these protections, it can't be because it's a reasonable person who uses syllogisms and is aware of the nearly infinite ends of life. By that measure we would be aborting persons from conception through the freshman or sophomore years of college. We're not making this judgment about the franchises and rights that attach to human beings in a strictly empirical way."

At this point Barry Schwartz, a psychologist from Swarthmore College, entered the discussion, challenging Arkes's earlier assertion that technological changes don't alter the domain of moral judgment. "Technology plays a much more fundamental role," he said, "and potentially a positive role. Because technology, not technocrats, can help us refine the categories that we use to understand ourselves and other people. One can make distinctions—at least in principle—about what human organisms are capable of doing at different stages in their development, prior to and after birth, and those distinctions should have moral consequences. The problem with technology as it is currently practiced is the technologist's insistence that we keep technical progress completely divorced from moral reflection. But that's nothing that's intrinsic to technology. We can have technology that helps us refine our understandings of when we become persons, what makes us persons."

"Technology can do that?" Berger asked doubtfully.

"Technology properly informed by moral reflection can do that," Schwartz replied.

In the exchange that followed, Schwartz noted that all cultures assume the right to impose the will of the culture on individuals. Arkes insisted that a process of reasoning lies behind such a cultural imposition.

Neuhaus asked whether or not the discussion was excessively complexifying Arkes's point. Oversimplified, isn't it simply the point that some rational justification, based on moral principles, lies behind cultural rules?

But Schwartz asked, "What's wrong with the justification 'I like it this way'? That, according to some, is the justification by which most people in modern industrial cultures now lead their lives. They don't regard themselves as moral beings at all."

In a final contribution to the morning's discussion, Robert Destro of the Catholic University School of Law, who is also a member of the U.S. Civil Rights Commission, connected these issues with the question that was to become the focus of the next session: "Don't some of these questions go back to the very basic question of society versus the individual, to the question of who is a member of society? A member of a society has certain duties and obligations, but in American law we tend to speak only of rights. If we go back to the questions of a mother's duty to her child or a husband's duty to his wife, American law tends to abrogate that completely. Divorce provides an example: in its development through Supreme Court decisions, marriage doesn't even get the same treatment that a common contract would. It's an attack on the principle of duty as the foundation of a society. So we allow the technocrats to come up with the notion of pre-embryo, or the more generalized Kantian notion that a member of society is one who is able to reason. Isn't Dr. Arkes's paper really about who's a member of society, whether or not we call that a person, and what obligations flow from that? I'm referring to the basic notion of this whole thing we call society, people banding together, the precursor of the polity. The degree to which a person is entitled to protection is dependent on that person's being part of that community of people."

In his concluding remarks, Arkes agreed that there are duties and obligations, but doubted that the wrongness of taking the life of a child in the womb is dependent on the membership of the mother in a particular society. While it might be another way of getting at the problem, he made it clear that it was not what he was saying. He closed with a reaffirmation of the importance of universal or categorical principle.

POLITICS AND MORALITY: THE DEADLOCK OF LEFT AND RIGHT

Following the opening session's debate, which set universal principles against a contractarian basis for moral judgments, the focus of the second session shifted to the contemporary cultural context in which such a debate takes place, and the question of whether or not the viewpoints of both the Left and the Right are deeply affected by the individualism of the liberal ethos. The underlying disagreement over universal principles remained in play, however.

In his introductory remarks, Christopher Lasch of the history department of the University of Rochester reflected on three factors that lay behind the writing of his paper, which now became the subject of discussion. First, he emphasized the sense of political deadlock between the Left and the Right, in which the conventional categories no longer seem to provide clarification. Both sides have adopted the same individualistic conception of rights, based on a detached, autonomous, morally choosing self. Such thinking underlies both the Kantian and the utilitarian positions in American political culture, he suggested. Both sides share the same belief in the desirability and inevitability of technical and economic development, and the same tendency to allow technological and economic change to set the political agenda and define the debate. Raising a question about why abortion had become such a central issue in the present conference, he labeled it only one of many such issues but an illustration of the way in which both sides cast their arguments in terms of rights. Second, Lasch asked what happens when the discussion shifts from a question of individual choices to a focus on the pattern of character in societal terms? Third, he raised a question about what constitutes "the good life." The public debate simply assumes no agreement on this issue and wrangles over specific moral choices. How much moral consensus on the good life does a society require? he wondered.

He noted that the conclusion of his paper points to a "fraternal" conception of politics as distinct from both the individualistic and the communitarian concepts presently operative. Conceding that he had not fully worked out this third way in his paper, he identified it as a fraternalism rooted in a particular cultural tradition rather than a mystic notion of universal brotherhood. It is not a sectarian model of brotherhood but one grounded in a common past and in the mutual demands of the present.

In opening the debate, Neuhaus expressed appreciation for the way in which the paper reminds its readers of how thoroughly liberal the American context is. Responding to Lasch's question about why abortion had become—and was likely to remain—a central issue for the discussion, he referred to *Roe v. Wade.* "In no other instance has a Supreme Court decision of great moral moment been made with the explicit denial that it is a moral decision."

Character, Fraternity, and Community

Stanley Harakas, addressing Lasch's call for emphasis on character and fraternity rather than individual choice, spoke first. "I understand you to use 'character' to appeal to a fundamental, bedrock moral embodiment. The assumption is that when we talk about character, we know what it is." He probed this assumption. "Character reflects the value system, the ideals, the worldview of a society. The appeal to character includes an appeal to a worldview, finally. Fraternity reflects a more organic understanding of who we are as human beings. I wonder how those two ideas connect."

Neuhaus deepened Harakas's point: "You're saying character is an empty term unless it's given moral definition by some communal worldview."

Lasch answered, "Character can refer to a fundamental disposition toward the world that is educated by experience, including the experience of making moral choices. It doesn't necessarily have external or inorganic implications." Referring to the related concept of "identity," he noted that while earlier the term had suggested continuity of personality over time, more recently it has come to mean the opposite, "something so transient that it is no longer possible to speak of its continuity."

Harakas said to Lasch, "I wonder if I might tell you the answer I was hoping to get from you? You posit something like fraternity, and I read that as something more than just voluntary association. There is something fundamental about being a human being, part of the human family, that relates you to everybody else in a way that creates a moral and spiritual bond. Being a human being in community means and imposes certain things. Then we can talk about character as an alternative state to individualism. If we assume that what's first is the corporate reality of our humanness, that assumption works itself out in how people develop in relationship to one another. I think the model of 'family' is precisely that—here I don't mean the nuclear family but the idea that we are family as human beings—and that means we are particular kinds of people if we fit properly into the family."

Lasch objected to the universality of this concept. "Communities are very specific and particular. That's what makes them communities. 'Universal brotherhood' didn't prevent Southerners from enslaving blacks. It's a very weak reed when it comes to moral power."

"But without that sense, there cannot be the smaller community," Harakas responded. The exchange continued briefly until Neuhaus concluded it with the observation that the two forms of community, the parochial and the universal, may of necessity be complementary.

Lasch, however, strongly reaffirmed what had become the

communitarian or fraternal side of the continuing debate over the basis of moral judgments: "I am far happier with moral prohibitions founded on a sense of 'We don't do that here' than on some appeal to universal principles."

Sidney Callahan added another dimension to the concept of community. "I think the way Dr. Lasch has worked in trying to define community is very important: common task, common heritage, common parents, common brothers. I would also suggest that some important things come out of studies of infants and mothers that show the way a community develops between people who cannot talk, who do not yet have language. In that kind of community, participants have a common attention (they look at something together), a common project, and common feelings. And then, of course, there's the fact that community develops because people live in the same common place in space. My definition of community is that you know people you didn't choose to know; they're just there. The question, though, is how much family resemblance and core cognitive overlap are necessary? I think it would be necessary to have a common core to make the rest of the project go."

"To have community in which you can meaningfully talk about character?" Neuhaus asked.

"Yes, and to make it work," Callahan replied.

Lasch answered by stressing the shared history that makes community. "This also carries with it enough to make it possible to argue with each other. That's all the agreement that's needed: a common language is really all it comes down to. I mean that not just in the literal sense of the term; I'm including all the associations involved."

"What would make people want to argue with each other?" asked Neuhaus. "Why would they be taking each other seriously enough to argue? Because they have a vested interest in the outcome of the argument?"

"Precisely," responded Lasch. "Because they have a vested interest, and perhaps a larger vested interest than is usually apparent in legal battles. There are issues that make it more difficult than it used to be to avoid the questions raised. Tomorrow we'll hear about ecology, which is bound up with all the technological issues we've been talking about, too. When we confront the long-term results of the technological interventions into nature that we've lived with now for a long time, we have an issue that's harder to assimilate into the old framework—partly because injuries done to nature are not easily talked about in terms of rights. In this case, the weighing of rights doesn't answer the question."

The Ubiquity of Liberalism

Richard Stith entered the discussion at this point, affirming the Lasch paper's emphasis on the ubiquity of liberalism: "It seems to me that just about everybody is basically a deep liberal nowadays, and there aren't any counter-value positions being articulated. There isn't any danger of religious war resurging anyplace outside of the Middle East. This is kind of a bogeyman that Rorty, Rawls, and the hyper-liberals use to crush an opposition that isn't even there. Really, nobody's going to rebel anyway. And to me that's the tragedy of our times, and I don't think abortion is an exception; it actually demonstrates the tragedy. Both sides of the abortion issue argue from hyper-liberal principles based on individual rights. Nobody really asks whether or not fetuses are included among the individuals that we want to protect. And that's a significant issue. It shows nobody is trying to decide who is a member of the people that we have to be fair to. Nobody is trying to go beyond liberalism on the abortion issue. The upshot of that for me, politically, is that I don't have any problem in completely casting my lot with the communitarians, who argue for more moral legislation. I just wish the communitarians had a stronger program."

"Richard, I don't understand what you're saying," Neuhaus interjected. "It's certainly intriguing. But with all the issues that are being debated—abortion among them, a euthanasia debate in California, all the issues that have brought us here—people feel impassioned about them. Look at the Bork nomination last year. Wasn't that an impassioned rallying of forces on all sides? These wars weren't explicitly religious in character, but certainly they had religious intensity."

"It's a kind of mopping-up operation," Stith replied. "Little remnants of contrary views of what makes the good life are still extant. And then there's the family, which is sort of an anomaly for bourgeois individualism. There are these little remnants, but they aren't really changing things. I say this after having spent a year in India, hoping to discover a different view of the nature of man developing there, and not finding it. After the 1960s there were all these hopes that Maoist China was coming up with a different point of view. But nothing is really left of that. They're all little side issues, trivial remnants."

"Why do you say the pro-life position is operating on liberalism's premise?" Neuhaus asked. "Why is it part of the liberal triumph?"

"Individual rights. Because if babies are people, they should be given the same rights as other people."

Here Robert Destro interjected a comment. "I have a problem with what Richard Neuhaus said. I looked at the polemics over the Bork nomination this way: the dominant theme was that the individual doesn't impose his or her views on anybody else because that's the normative moral value that we have all accepted. And the sense about Bork was that he was going to do that. The Reagan administration didn't respond at all; it didn't go after that argument."

"But look at Bork's defense—that he stands for *not* imposing personal views," Stith answered. "He's operating out of the same amoral liberal theory as his opponents. There isn't a deep difference between them, just a surface difference."

Neuhaus summarized the exchange. "Richard, you're saying—and I gather there's a lot of agreement with your point—that within our pervasively liberal context even those who view themselves as being opposed to liberalism are operating on its premise."

The Relativism of the Liberal Ethos

Stephen Post then related the same issue—the pervasiveness of liberal culture—to the field of education: "I think education is a real problem. The difficulty is that when we teach ethics, we are confronted with the psychological impact of cultural relativism. There are issues—like abortion—that for some are very difficult to resolve. Students throw up their hands and say, 'What's good for her is good for her; what's good for me is good for me. We can't have a conversation.' That attitude is simply amplified by the kind of texts we use in teaching ethics courses in colleges and universities, which give little snippets—five pages of such-and-such a position, five pages of the other position. Education is at fault for putting its imprimatur on the kind of relativism we're trying to get away from. Liberals are concerned about 'indoctrination'—they use that term. Indoctrination is a problem, but no one wants to distinguish indoctrination (or totalitarianism) from absolute moral principles. I think there are many absolute moral principles that we would agree on, such as 'do no harm,' veracity, justice, and so on."

Christopher Lasch intervened: "I'm not sure you could really argue that the schools fail to uphold those absolute values you just named. But the trouble with absolute moral principles that we can all agree on is that they're so boring and bland. On the one hand, nobody challenges them, but on the other hand, they don't give people any real knowledge of the moral world or any inner resources to deal with moral problems. Those principles are things we all know, but precisely because they are unproblematic, they're also not very interesting or helpful."

Post objected: "But they *are* interesting. And we do become better people if we bring to consciousness and articulate these basic principles. Granted, things get difficult in the application of principles because of empirical or even metaphysical or theological disputes. But the umbrella is essential to any kind of moral fabric."

Lasch drew the line sharply with his rejoinder. "Well, I just don't have that much confidence in principles. And for that reason, I don't see the problem of moral relativism as serious, either. It seems to me that it appears serious only to people who believe that morality begins and ends with the articulation of these great, universal, absolute principles."

Francis Canavan called the pervasive moral relativism "the unofficial public philosophy of the United States. But if you push hard enough and ask enough questions," he continued, "you'll find that underneath there is a bottom line. I don't mean that we should throw up our hands in despair. On the other hand, I don't think we should underrate what Barry Schwartz was saying this morning—that in wide circles in this country, 'We like it this way' is the ultimate answer, and there's nothing more to say. That has packed into it a set of moral absolutes: 'Because we like it this way, you have no right to interfere and therefore an obligation not to interfere.' Can a community be morally neutral? I say 'Clearly, no.' No community is morally neutral, and the pretense that says so is flimflam, and we shouldn't be taken in by it. This ought to be challenged and challenged relentlessly. To bring it into debate is not to predict the answer; maybe the answer will go against us. Nonetheless, questions like this can and should be discussed, and we should never be satisfied with a put-down or 'You're imposing your values.' Of course I am. But then I ask, 'What are yours?'

"This gets back to community," Canavan continued. "Earlier someone defined community as a group that is bound by reasoned discourse. I think the society is in trouble if it's not willing to engage in such discourse. Since the Enlightenment, reason is increasingly reducible to mathematics, which means that contemporary reason is incapable of making moral judgments. This too should be put on the table and discussed."

Neuhaus pointed out the connection between Canavan's point and Lasch's comments in his paper about unwillingness to give offense.

As the discussion continued, Barry Schwartz picked up the question of the culture's pervasive relativism. It has its source, he said, in "what is almost certainly the dominant institution in all of our lives, the market. The official ideology of this country is the ideology of the market, which invites relativism. Who is to say at any given time what is appropriately included in the market and what is appropriately withheld from the market? That's up for grabs. As long as we restrict our concerns about the good life to things like abortion and birth control and don't ask questions about what is appropriately up for grabs in the market—what is presently in the market that shouldn't be—I think we will be forced to face this kind of commitment to relativism in the future. Look at our discussion thus far about guaranteeing the *good* life. We spent most of the morning just talking about guaranteeing life—that is to say, taking a view on abortion—without worrying about what commitment we have to guaranteeing that the life be good. I don't think that we as a nation have much of a commitment, in our economic institutions, to seeing to it that the lives some of us think ought to be allowed to come into being are *good* lives. If we're committed as a community to guaranteeing or even fostering the good life, then that surely extends to considerations of economic distribution that at the moment are ruled out of discussion by the very same people who insist that we have to take a position on abortion and birth control. There's an inconsistency here. And I think the most important point in Lasch's paper is this notion that almost everyone is incoherent when it comes to economic views on the one hand and moral ones on the other. It doesn't matter whether we're talking about an economic conservative or a cultural liberal—both have incoherent positions. That's the point we need to be addressing."

Randomness as the Basis of Equality

Jean Elshtain, entering the conversation, made a connection between an issue discussed in the morning session and Lasch's comments on the pervasiveness of the language of rights. "This morning when Dr. Nathanson mentioned interventions in pregnancies, it occurred to me that a person was missing in the discussion— the mother. The fetus exists in an exquisitely, complicatedly social and embodied environment from the very beginning. It already has a social context with social connections that radiate out, so to speak, from the woman's body into the wider community. Nathanson talked about whether or not it's okay to cryogenically freeze embryos and separate them. Well, if one is working with a strict language of fetal rights, as long as fetuses aren't killed, presumably that would be morally acceptable. But the image that flooded my mind was of separation, of loss, of being ripped out of one's first and arguably most important human environment—the terrible violation of our human nature that that involves. There's something there we could work with to try to get out of the abstractness of some of these arguments about rights and first principles and so on. We could think about how we are essentially embodied human social selves from this very first moment.

"I have one other point to make. If we begin to intervene, as we're already doing, in the process of reproduction in ways that target certain categories for selective elimination—whether it's eliminating the defective fetus or choosing one sex over the other—we undermine the basis for equality, in the political and moral sense, at its root. We begin to say, 'Those sorts of beings ought not to be here,' and our obligation and responsibility toward them are subsequently diminished. The natural lottery— the throw of the ontological dice, if you will—is the basis for equality. Randomness is essential to equality, and if you undermine that, you undermine equality.

"That natural lottery is the basis for the kind of community that Kit Lasch is talking about. You don't say, 'I think I'd like a community where the people are all six feet tall and perfectly divided between males and females, and there aren't any handicapped people because they're a drag.' "

Some discussion followed, and Neuhaus concluded the exchange: "Jean, I think you're saying something that is elegantly important, which is that the whole thrust of the universal (which gets us back to a lot of other things that Kit also started us thinking and talking about) is, oddly enough, premised upon the contingent. We're concerned about equality as a moral good because there is inequality. And if, indeed, we have control over those contingencies—those accidents—then the moral burden for working for equality is removed, because we've already exercised our moral choice in designing the product."

Elshtain agreed with this point, and Noel Reynolds then interjected a comment. "This could be read backwards, of course, to draw a very pessimistic conclusion about the idea of equality. If in fact our commitment to equality depends on our not engaging in these kinds of things, that seems to me to be a very shallow and endangered premise for that commitment. There's another kind of equality that I haven't heard mentioned here—the equality that comes from man's relationship to God—which wouldn't be subject to that kind of qualification."

"I think that would be undermined as well," Elshtain replied.

In the same vein, Neuhaus said to Reynolds, "If God is displeased, or if those things that are ordinarily thought to be the consequence of God's will were displaced by human control, I wonder if it wouldn't undermine that premise of equality as well?"

Here Robert Destro made a comment, connecting the discussion to a point in the Lasch paper about the danger of leaving divisive issues with technological dimensions out of politics and treating them as technical questions that should be left to experts. "I wonder if that's not what you're talking about here, that equality becomes something that you leave to experts. I can offer a personal example here. One of my colleagues on the Civil Rights

Commission says, 'Equality is the management of diversity.' If that is said, then someone else becomes the expert at managing diversity."

The Inability to Live with Disappointment

After agreeing that the comparison was relevant, Jonathan Imber of the sociology department of Wellesley College returned to the question of technology and morality. "It strikes me that the technological aspects of these discussions are ineluctably filled with moral dimensions," he said. "People make decisions about which technologies will be used and not used. In our society this entire logic of the use of technology has led toward a greater reliance on controlling images about how we ought to live. When we talk about selective abortion or any of the other particular technologies that can guarantee a certain outcome, we are enlisting the aid of a logic that has been called 'screening.' Where does that expectation for guaranteed results come from? I think it goes to the heart of the culture we're living in right now: most people don't like disappointment. Our embracing of technology has become an alternative to dealing with disappointment. It was once a function both of faith communities and of the professions to suggest that disappointment is something one sometimes has to face."

Imber went on to suggest that the medical profession was largely responsible both for the laws struck down by *Roe v. Wade* and for the conditions that led to *Roe v. Wade*. The change came about, he said, because of the demands being made on the profession—"that no one can be disappointed, that no one can be without a guarantee about what the outcome of a procedure or a treatment will be. One of the reasons people don't want to make judgments about other people with regard to decisions on abortion is that no matter what the decision is and no matter what the reasons for it are, an element of disappointment and pain is always involved." He indicated that moral quandaries still surround the decision to have an abortion, which is why it is not yet accepted as simply another form of birth control.

"The way in which that particular decision is expressed," Imber noted, "is related to the way in which the service is provided in our society—namely, by physicians." He suggested that from the standpoint of the transformation of the medical profession and the provision of abortion services, the companion case, *Doe v. Bolton*, was more important. That decision said that the law could not force a physician to perform an abortion, but that if an abortion were to be performed, it had to be performed by a physician. Subsequently among physicians there was an easy accommodation to those principles.

"The result is that approximately 250 clinics in the United States perform somewhere between 70 and 80 percent of all abortions, 90 percent of which are in the first trimester. This suggests that abortion *has* become a routine form of birth control, but it doesn't suggest thereby that people who resort to it or physicians who perform it are without an experience of moral complication."

At the end of these remarks, Neuhaus noted that Imber has written a fine book entitled *Abortion and the Private Practice of Medicine.*

Special Duties and Relationships

At this point Fred Sommers returned the discussion to the contrast in Lasch's paper between the particularist communitarian approach to moral judgments and the universalistic-individualistic approach to moral judgments. This is associated with another contrast between evaluating the deeds (virtue-based) and evaluating the person (duty-based). "Much of contemporary ethics is universalistic," he said. "The 'special duties,' usually kinship-based and particularistic, are often omitted. That's why I liked Dr. Lasch's notion of character. But I would like to tie it to social role—that is, character depends on one's position in one's family, one's country, and so on. Dr. Elshtain mentioned the mother who owed a special obligation to the fetus and who failed it. All this is omitted when we talk simply about human rights. If we talk about rights at all, we should talk about the rights of this fetus against the mother, not against the whole society but against the mother or the father. These two people authored the fetus, after all." Sommers concluded by saying to Lasch, "That's the kind of thing that comes out of using your approach, which I like very much."

Lasch noted that "the 'special duty' that I think is in need of rehabilitation is the whole question of duties associated with

citizenship." He also noted that thinking this way is deeply opposed to the dominant trend in our culture of "not wanting to be
imprisoned in a social role, which is central to the whole rhetoric
of rights and liberation."

Stanley Hauerwas of the Divinity School of Duke University commented on "the language Professor Sommers used,
which I quite appreciate but want to dissent from—the language
of 'special relations.' Why should I call friendship or the relationship between myself and my son a 'special relation'? The only
reason you do so is that you presuppose that the relationship between strangers, which you rest on grounds of impartiality, is the
determinative relationship, from which you have to give justification for something called 'special relations.' That's already not
thinking rightly. It's already to be corrupted. Aristotle devoted
two books of his *Ethics* to friendship. Friendship works as the crucial moral relationship that pulls us into the moral life."

Hauerwas went on to discuss the whole problem of how
liberalism gives an account of patriotism. "Patriotism becomes irrational on liberal grounds exactly because boundaries of nations
are always arbitrary, since the principles that liberal societies hold
should be the principles held by anyone. That's the reason
liberalism is always imperialistic, and the market becomes the
form of imperialism which the army then follows. Patriotism
makes sense because it isn't a universal. It is something I must
cherish in terms of its historical particularity, and therefore I do
not appeal to some further universal to justify it. Issues of boundary and history, the terms by which Kit Lasch was suggesting
community is defined, become absolutely essential."

A Constructive Alternative

After a brief exchange on the nature of community, in which Hadley Arkes reiterated his earlier insistence on principle rather than
tribal understanding as the basis of community or nationhood,
Thomas Oden of Drew University Theological School entered the
conversation by addressing Christopher Lasch. "I share much of
your critique of the conditions of moral discourse in our time,"
Oden began, "but I'm looking for a constructive alternative. I
want to hear more about the politics of fraternity. You approach
it with the notion that brothers must live with each other. They

have to get along. They have common parents. But I don't see where that leads us. There is much about the concept that is fascinating, but I'm looking for the constructive alternative.

"One more comment. You describe as diametrically opposed and polarized the pluralism of the Left and the communitarianism of the Right. These are viewpoints that seem to me to be much more overlapping and mixed. At my university I plead for a certain kind of pluralism all the time, because there seems to be a liberal consensus against which I have to struggle. To me this picture seems much more diffuse and gray than black and white, which is how you portray it."

Oden then made a final comment on Lasch's reference to disappointment and suffering and the technological interventions that seek to circumvent any kind of pain. He suggested that in light of the Jewish and Christian notions of redemptive suffering, this might become an important part of the conversation.

In responding to Oden's last point first, Lasch said the issue was the absence of any point of view that sees value in suffering, which is not seen as part of "the good life." (This became an important issue later, particularly during the second meeting of the conference.)

Foregoing the opportunity to respond to Oden's challenge to his description of the polarization of the Left and the Right, Lasch moved on to the question of a constructive alternative: "It seems to me that that's what we've been talking about. It's the difference between two ways of thinking about a lot of issues. Does morality rest on principles or on a shared experience of community? Is community defined principally by agreement or by common history, as I would argue?" Referring to the issue of slavery in the Civil War, Lasch suggested that Lincoln operated on the fraternal basis. "He illustrated the contrast between an absolute principle (the principle of equality) and a kind of historical intuition that led him to think that the preservation of the Union, of this community, flawed as it was, was more important."

Neuhaus noted that it was by no means incidental that the question of abortion and the analogies with the Civil War and slavery kept arising. "That's the other great experience in American history in which this 'moral overload' of our procedural polity occurred, and at that point the polity couldn't be restored without bloody civil warfare. We hope that's not going to be the case this time around."

Neuhaus went on to summarize the course of the discussion thus far: "I see a very interesting, coherent incremental development in the conversation. This morning Hadley Arkes gave us a way and an argument for a way of addressing these questions. He recognizes fully the difficulties that our culture and its liberal premises pose; nevertheless, that's the way he would like to see these questions addressed. Obviously not everybody was entirely persuaded by Hadley's argument. Then, this afternoon, Kit Lasch made this much more concrete in terms of the historical, social specificity of the structure of such debates in our kind of society. I'm sure he would be the first to agree with Tom Oden that his paper is longer on the analysis of what's wrong than on a convincing, comprehensive alternative proposal. But that's why all of us are here: to help construct that convincing, comprehensive alternative."

Fraternity and Individualism

At this point Brigitte Berger re-entered the discussion, identifying the issues of the discussion as the individualistic conception of society and the communitarian conception of society. Lasch, she said, "suggests the notion of fraternity as a way out of this impasse. And I would urge him to look longer at this fraternal notion, particularly if he is concerned with the formation of character. I would propose that the fraternal notion and the notion of character are not compatible. Of all the familial images that might be chosen, the fraternal one is the weakest. It's the one that doesn't involve the formation of character. It's a much more competitive notion, a very different one. But what is character formation? It comes out of the imprinting of certain kinds of values and morals that are poured into children. So I have a lot of problems with the fraternal notion. If you want to use a family image, it should be the parental one."

With regard to individualism, Berger said, "All of us know that individualism is a peculiarly Western preoccupation. It is part and parcel of the bourgeois culture, which has come into existence during the course of the extension of the market to an ever increasing number of people. The bourgeois liberty—that's what we're talking about. In his paper I think Lasch slightly misunderstands the interest that Peter Berger and I have in overcoming the communitarian and the individualistic principle. For us the issue is that the only way of getting out of what I would call a hyper-individualism—which we are seeing run rampant in our society today—is through a religious connection. It's the religious reference to something that's part of you, that doesn't come out of the collective life. During the last year and a half I've studied urban migration in the Third World, how people in the slums order their lives. What strikes me is that, while the Catholic bishops call for 'planning for the poor,' the poor have planned for themselves. How? Through morality coming out of the experience of family values. By necessity they are driven into community-building. And by necessity they are driven into asking the questions of the higher principle, which again is a religious principle."

Lasch responded to Berger's suggestion that fraternalism is the wrong model for society, particularly her suggestion that fraternalism is competitive rather than character-forming. "Well, of course, it's both. Competition itself is a highly character-forming activity, central to those practices in which virtues are embodied in the course of education. The best form of competition is argument, when the antagonists take things very seriously. It seems to me that there is an irreducible tendency to try to remake one's siblings. The joint history of siblings—which, of course, presupposes the history of parents and grandparents (this seems to me another advantage of fraternal imagery)—is a lifelong argument in which they try fiercely to change each other's minds. In the process they form each other's characters."

At this point Father James Burtchaell joined the conversation with further reflection on the individualism of the liberal ethos. "There is hardly a major psychological disorder in the Western world today that isn't somehow associated with the way in which people miss belonging. Being left as 'autonomous individuals' has been a most destructive experience." Teenage pregnancies, truancy, child abuse, and many infantile psychoses are related to children who grew up not belonging, Burtchaell said, as is the high rate of depression among the elderly who have been given "autonomy." This pathology cannot continue indefinitely, he suggested, without shaking the confidence in individualism.

After questioning the commitment of Enlightenment

liberalism to utter equality, Burtchaell concluded by seconding Berger's request for enlargement on the fraternity coda in Lasch's paper. Noting that the family is the closest of all communities— knit of a bond of promise (husband to wife and wife to husband) as well as a bond of blood (parents to children and children back to parents)—he pointed out the interdependency of the two bonds. "In talking just about community and special duties, peculiar obligations, we tend to dim the old notion of fidelity and commitment and steadfast belonging. In the bond of promise, one marries someone who is only slightly less a stranger than the children will be when they arrive. If taken seriously as a lifelong promise, that bond has no rival as a model about how we are to deal with one another in love."

Community Based on Law

Recalling the various models of community that had been intro-duced into the discussion—community based on historical acci-dent, on shared principles, and on the family—Noel Reynolds ob-served that there are communities of all these kinds. All have validity, but ours is a society based on law. "A society of law has a different central function," he said. "It is to make possible the peaceful coordination of activities between strangers. In this kind of coordination there is no assumption that all share the same principles, but there is an assumption of a widespread commit-ment to *some* principles that would be minimally necessary to make such a system of law work. And certainly there is a commit-ment to the same legal procedures. So in a community thus defined there can be moral diversity but legal unity. That seems to indicate why the ethic of tolerance turns out to be so central. The mistake, referred to here so often, is not tolerance; that ob-viously isn't a mistake. The mistake is the failure to see that what is important is not waging rational argument to identify truth that can be implemented by judges or the state, but rather recogniz-ing the importance of rational persuasion in developing consen-sus and resolving social issues. When people vote, we don't know if they're voting on practical grounds or in response to a bribe or a moral conviction. So in a legal society we allow all those things (though not bribery, of course) to play a role. The thing that mat-ters finally is agreement, and appealing to moral principle is one of the ways we have of trying to reach agreement. That's the relevance of moral principle."

As the session drew to a close, Stanley Hauerwas connected technology with the discussion of liberalism generated by Lasch's paper. "We don't want to be bound by the limits of our procreative possibilities, which are God-given. And what people are trying to do is give Gnostic reasons to explain why they're having children in the first place, in terms of perfectionistic accounts, which ultimately means that the product must be destroyed when it doesn't turn out to meet those perfectionistic standards. So I think there is a deep relationship between the problems Kit Lasch has raised about the political ethos we inherit and the technological challenge before us, because we don't have any way even to understand the issues other than through the language we have been offered by liberalism. And any other alternative in the end sounds, to liberal ears, like fascist romanticism."

"This requires control, technologically or otherwise, over the contingencies that get in the way of our universal self-delusions," said Neuhaus. Hauerwas agreed, and that ended the general discussion.

THE NEW EUGENICS AND FEMINIST QUANDARIES

In the third session of the conference, the emphasis shifted from the philosophical basis and the liberal context of moral judgments, treated in the earlier papers, to focus more directly on technological interventions in human life. Highlighted particularly were those interventions related to women and reproduction, as treated in terms of the feminist movement in the paper by Jean Bethke Elshtain.

Elshtain summarized three ways in which feminists have positioned themselves on what this conference is calling the new eugenics: the radical non-interventionists, who reject all technologies as patriarchal and designed to oppress women; the radical pro-interventionists, who welcome all technologies that free women and who, in the extreme, assert that women will be fully free only when technology reaches the ultimate stage at which reproduction can take place entirely outside the female body; and the moderate pro-interventionists, who favor some forms of technological engineering and reject others, the standard being what is "good for women." Elshtain positioned all three groups in the wider frame of "ultra-liberalism," with its emphasis on rights and the radical autonomy of the self.

She added that science can encode a structure of violence. "The new technologies are new in that they're being applied to human beings. They've been applied in animal experimentation for a long time. There's a sort of moral *cordon sanitaire* around these efforts: as long as such experimentation involves animals, we aren't too worried about it. We don't absorb these experimental subjects into the moral universe. But what at first was limited to animals is now being done with people. It's quite clear from the literature produced by the reproductive technologists themselves that they see themselves as experimenting with the female body. Violation of the body in search of scientized production and control is a matter that we need to think about and concern ourselves with."

In opening the discussion, Neuhaus pointed to the logical flow of the conference thus far. "Hadley Arkes helped us in thinking about how we make moral judgments with respect to many things, including the contingencies of technological change. Then Kit Lasch moved us into a discussion of how that judgment-making is politically configured in American public discourse. And within that mix of public discourse, Jean moves us into the knottiest part of it, which is tied up with feminism. This is a particular instance of what Kit depicted more broadly. In some ways this is the most difficult—but also, perhaps, the most important—of the particular disputes within the context described by Kit Lasch. That's how I see the progression as we move along."

Rethinking a Theory of Society

After David Novak praised Elshtain's paper as an illustration of the way moral passion and intelligence can work well together, he said, "I've always regarded feminism—as I have come to know it but not to love it—as a late-blooming variety of liberalism's atomistic individualism. And I thought Elshtain's analysis was very penetrating. It laid bare what is wrong not only with feminism but also with the larger underlying theory of society out of which feminism emerged and in which feminism is intelligible. Do we have to hold on to the individualistic theory of liberalism in order to keep its achievements? I don't really think so. John Courtney Murray, for example, was trying to argue that Catholics could accept religious liberty, but on another basis. The fact that we like the results of certain historical developments which were based on a certain type of theory doesn't mean we've got to buy the theory. That's open for rethinking. And the rethinking will have to go beyond the theory of feminism to a theory of society. It will involve critique of the nominalist metaphysics that ultimately lie behind all this, and the dominance of the notion that science is the highest form of knowledge—that if we can't prove something scientifically, then we don't really know it. The rethinking is really necessary, because the lengths to which radical feminists go do become ridiculous—although they're not illogical, given where they start."

Neuhaus posed a question: "That rethinking is a project of generations, in terms of its becoming entrenched in law and popular culture and higher culture. Is all that very helpful, considering what we're facing? After all, that rethinking involves convincing, over the next few generations, the academic and intellectual elite in America that there should be a different foundation for the ordering of our society."

"Can it happen?" asked Brigitte Berger.

"I don't see why it can't happen," Francis Canavan answered. "Before the other theory came in, there was enormous resistance to it, and it didn't win because of its truth. But the point you make is that there are certainly a lot of immediate and practical things to be done in the meantime. It's true that societies don't live only on ideas. But the ideas do have to change. I suppose that in practice people stop when the line of reasoning arrives at conclusions that are obviously repugnant and ridiculous. At this point they say, 'Something's got to be wrong here. We really don't want to go this far.'"

"It strikes me that we couldn't eliminate rights if we wanted to," Elshtain responded, "and I take it no one in this room would want to eliminate rights. The question would be how we re-situate our understanding of rights and the subject of rights in such a way that we don't have what we now have, which is the understanding of rights in this atomistic worldview and its very particular ontology. That's a complicated issue. What we might want to reflect on is this question: What sources do we find in the society for renewal, for some alternative? It goes back to the concern that Richard Stith expressed yesterday—that maybe this is all so hopeless that we've just got lots of these mopping-up operations to undertake, and that a revolt is bound to be futile because the perspective is so dominant that there's no way to effectively resist it."

A Point of Repugnance

At this juncture Neuhaus highlighted an issue raised a few minutes before. "I hope we don't lose sight of the point that Frank Canavan raised. In the course of our discussion today it might be useful to get some other assessments of this matter of the point of repugnance, in terms of people saying, 'This far, no farther,' and what might trigger that. With regard to 'Baby M,' for example, there is some survey research which indicates that this issue has elicited enormous public repugnance. Similarly, where the abortion debate is concerned, the 1.5 million abortions performed per year is a statistic that is generally met with repugnance."

"But people don't want abortions to be made illegal," Elshtain noted.

"This, though, is the problem," Neuhaus responded. "Are they willing to pay the price in order to remedy it? And a whole host of questions is included: take some of the methods of reproductive technology, for example. In response many people will say, 'That's very frightening and disturbing,' but then, when they weigh such concerns against a person's 'right' to have a baby, they will say, 'We don't want to interfere with that.' So it's a conflicted repugnance, a repugnance that lacks remedy because the remedy would violate rights."

"I don't know where it becomes effective," Canavan interjected.

"Exactly," replied Neuhaus. "But I hope the question you raised receives attention."

Stanley Hauerwas picked up the issue. "The whole question of repugnance is really interesting. I just don't want to trust the common consciousness on that. The Germans weren't all that repulsed about what was happening to the Jews, although they might have been for a while. People might exhibit the right kinds of repugnance; they might not. They can get used to some unbelievable things."

Jonathan Imber, noting the sense of repugnance in media accounts of a recent case of infanticide, said, "The real difference is that the kind of infanticide which is practiced within the technological halls of our culture is a sanitary infanticide. It has been cleansed in a way that disposing of an infant in a garbage can cannot be. As long as we assume that the repugnance expressed at garbage-can infanticide is going to carry over to the hospital kind of infanticide, we're not really sensing the kind of controls these particular institutions have established."

"The medicalization of infanticide is going on," Elshtain agreed.

"This kind of control allows the sanitary horror to occur," Imber continued, "but it wouldn't for a moment sanction what this particular woman in the news accounts was accused of doing."

Neuhaus concurred: "That's extremely important, Jonathan—the sanitizing, and also the scientizing, of things that would otherwise be repugnant. Of course, in a lot of pro-life agitation, a great deal is made of the regular instances of finding discarded fetuses in the city dump. There's something weird about the use of these incidents. The people who are pointing them out know intuitively what Jonathan is saying—that somehow they can extract a response of repugnance by mentioning the trashheap disposal of fetuses, a response that doesn't occur when the fetuses are disposed of in some 'neat' way."

Richard Stith agreed with the point that a large number of people find abortion morally repugnant but wouldn't want to make it illegal. "Nevertheless, I would say that an overwhelming majority of people would want to make it illegal *to some degree*," he said.

Animals and People

Next Barry Schwartz entered the discussion with a reference to Elshtain's introductory remarks about the use of animal experimentation to develop a technology now applied to people. This, he said, has been going on for a long time.

"It's more frightening, though, if the experimentation isn't done on animals first," noted Elshtain.

"But using animals as models and developing a technology

which does things like curing cancer doesn't seem frightening," Schwartz countered. "There's nothing new about that. It's just being applied to a new domain that we regard as potentially frightening."

"On balance," answered Elshtain, "we may say that it's appropriate to rid the world of a terrible disease, and that we're morally prepared to sacrifice a chimpanzee to that, or whatever. But we might not concur with the moral attitude that the researcher has toward the animal—as inert matter, inferior material to which we have no moral responsibility. That attitude—to the extent that it becomes pervasive and then extends to defenseless human beings and maybe then to all human beings—may be something we wouldn't want to adopt."

Neuhaus interjected, "Isn't it also important to make the distinction that what has been going on for a long time is the use of animals for essentially therapeutic purposes? The new thing is to produce a better human product, as we have in the past produced a better animal product."

"But that's somewhat benign in context," Elshtain commented.

"We're also using animals to find out how people work," Schwartz said. "In moral terms there are all kinds of potential aims in the project of finding out how people work. For instance, there is the aim of making better people by improving the nutritional habits of pregnant women. Now that's an attempt to improve the breed, and one can come at that through all sorts of experimentation on the nutritive environment of animals. I'm not suggesting that this isn't a moral issue, but there's nothing new in it."

"I was suggesting that we should open the doors of the laboratory and reflect more than we have about what goes on in there," Elshtain responded. "It's too bad we're doing it only when we get uneasy about laboratory experimentation carrying over into human applications."

Surrogate Parenthood

Schwartz now moved on to another point. "A distinction has to be made between various kinds of acts and the justifications that people provide for them. The thing that struck me about 'Baby M' is that in all likelihood there would not have been much of a moral outcry if the matter hadn't been contested. The key thing about the case wasn't its contractual character. Another gloss that can be put on 'Baby M' is that it involved a socially concerned individual, Mrs. Whitehead, trying to perform an act of kindness for a couple who couldn't have a family, the Sterns. So this is an act of social connectedness that's not contractual in nature but rather reflects some sort of general shared sympathy and commitment that runs through the culture."

"You're not saying that about commercial surrogacy, are you?" asked Elshtain. "It's hard to say that it's an act of human connectedness, pure and simple, when the biological mother has a contract and is going to get $10,000."

"But even with commercial surrogacy," Schwartz responded, "it's not clear that the contractual nature of it is built into the act itself. When a woman decides to become pregnant to assist another human being, she incurs various inconveniences. Now one could construe the contract not primarily as a fee for services rendered but as recompense for expenses incurred. It's enough that the woman is willing to make this sacrifice for nine months of her life. Why should she also have to pay various costs? That's one way of interpreting the relation between the surrogate and the parent. When the controversy first developed in the public domain, I think many people tried to argue that that was in fact the intent. So, in many cases, surrogacy should or can be understood as an attempt to perform an act of kindness. I don't see anything that inheres in the act of surrogacy that makes it automatically contractual or commercial. Once Mrs. Whitehead decided she wanted to keep the baby, all these other problems arose, and the contract changed its character."

"There is the case of the South African grandmother who had a baby for her daughter," Neuhaus interposed. "In that instance no money was exchanged. And you're saying that in that case it could be construed as an act of connectedness, or charity."

"It would be very hard to interpret commercial surrogacy that way," Elshtain replied, "but that isn't the only kind of surrogate arrangement. Commercial surrogacy is something we don't agree on. I think the contract infects the entire activity. In cases of informal surrogacy, Schwartz's point might be arguable. But if you read the contract that Mary Beth Whitehead signed, you see that it was more than a fee-for-service arrangement. She was also required, at the request of William Stern, to abort if it was determined that the fetus was defective. She was to be given damages if she had a spontaneous abortion. The entire process of gestation was under contractual control and surveillance."

"This might turn into an argument about what the terms of such contracts ought to be rather than a categorical argument about surrogacy," Schwartz commented.

Elshtain rejected that possibility. "I would simply stop that argument. I would legislate against it. That's what I would do."

At this point Hadley Arkes interjected a question: "Would you object to it if there were no contract involved, Jean?"

"It still makes me a little queasy to think about," she answered. "What people decide to do informally shouldn't be under full control of the state. But morally, it's tougher. There was a case in which one sister gave birth for another sister who was infertile, and there was no money exchanged. I know that I'm uneasy when I read about such things, and I haven't fully explored why. But commercial surrogacy is simply out. I would outlaw it."

Schwartz ended the exchange with this comment: "It's not clear to me at any rate what defines commercial surrogacy. I don't think it's sufficient to say that if there's a contract, it's commercial, and if there isn't a contract, it's not. Contracts can be written only as guarantees about worst cases."

The Property Issue

Robert Destro offered a legal point. "The Baby M case was portrayed as a fight between Mr. Stern and Mary Beth Whitehead when in reality the exchange was between Mrs. Stern and Mrs. Whitehead. Mr. Stern already had the rights of a father, so the legal exchange was that Mrs. Whitehead had sold her baby to Mrs. Stern. Thus the question really is about property. Doesn't the commercialism-or-non-commercialism issue turn on the baby, or the womb, as property? In property law, whatever the reason, it would be considered an exchange."

Elshtain agreed. "Yes, I think it's baby-selling. That's finally where the point of moral uneasiness comes in. It also establishes an absolute ontological equivalence between masturbating into a tube and carrying a child for nine months. In fact, the weight of the contract is that the donor of the sperm gets to claim the baby as his. The mother, therefore, has absolutely no rights—it's hard to get away from the language of rights—and no claim. In that overwrought conversation that was played in the court, Mr. Stern tells Mrs. Whitehead that he's the one who has the proprietary right over the child."

Schwartz pursued the theme of proprietorship. "Is there something wrong with women talking in terms of ownership and control of their bodies? If they don't own their bodies, who does?"

"Bodies aren't owned," was the reply.

"That's what I assumed the response would be," Schwartz said. "But at issue here are the rights and responsibilities that ownership of *anything* entails. There are things that are owned, and it doesn't follow that one can do anything whatsoever with them. If we move the language of ownership from the domain where it normally operates to the domain of human beings, we misuse it."

"But you agree, Barry, that bodies are not owned?" Neuhaus asked.

"I don't think the problem is applying the language of ownership to bodies. The problem lies in what we understand ownership to allow us to do. I don't think that people who own companies should be allowed to do the things that they are currently allowed to do. That's the source of the problem. We have a misguided notion of ownership as it applies to things, and when we apply that notion to bodies, we come to the wrong conclusion."

"That leaves open the possibility that there would be a right way to use the language of ownership," noted Elshtain.

"That's right," replied Schwartz.

"That's where I would say no," Elshtain insisted.

"There is a clear religious answer," Neuhaus pointed out. "It claims that you are not your own but that you are bought with a price. That's the language of ownership, and it very clearly specifies who the owner is."

Moral Judgments on Related Issues

Hadley Arkes directed the discussion toward political decisions that he predicted would be coming up soon. He said to Elshtain, "We see that already foreshadowed in California in the Watkins case, which insists that classifications based on gender in adoption and marriage are simply insignificant. And we saw that in the presidential-primary candidates who were quite willing to sign on to the notion that gays should have the right to adopt children. Would your argument yield up a ground for judgment on those questions?"

Elshtain replied, "The form my argumentation takes is different from the direction you usually go, in part because you, coming out of the constitutional law tradition, are always looking for grounds upon which to make decisive judgments. There are some practices—for instance, commercial surrogacy—that I don't hesitate to argue should be banned absolutely. Then there are areas such as non-commercial surrogacy that make me morally queasy. It seems to me one can't move in a prohibitionist direction with reference to those kinds of issues."

"Can you make a moral judgment about non-commercial surrogacy?" Neuhaus asked.

"You can raise some moral questions about it, questions that would and should invite reflection," Elshtain responded. "But what makes it complicated is that the people who are engaging in it see it as something done out of a sense of caring and obligation tied to other human beings. They're making the ultra-liberal kind of argument that I blast at the beginning of my paper. The child is seen as a gift, something offered out of love. I would want to ask questions and engage in discussion. I would see a persuasionist strategy there."

"Hadley wants a principle," Neuhaus observed.

"He wants a necessary universal axiom!" Elshtain agreed.

"No," said Arkes. "I wasn't trying to push you into the cast of my argument. We're both political scientists. We're both aware of the political arena in which these questions have to be discussed, and decisions are going to be made, willy-nilly. How strong is your insistence from an argument that seems to be grounded in nature; how strongly do you regard your arguments on surrogacy?"

"Let me resist your characterization of my argument as grounded in nature," Elshtain countered. "It's grounded in an awareness of our essential embodiment and in the reality of the moral imperative embedded in that. It's a moral understanding of our human identity. It's not the usual argument from nature."

"Another kind of question along these lines," Stanley Hauerwas put in, "is this: What reason would you give as to why we might not engage in extensive research to enable men to carry children?"

To this Elshtain replied, "There you have a very strong argument about the violation of the created being that that involves. That possibility deepens our hubris and our arrogance about control and the notion that human beings are creatures that we can work over in order to get the kind of ideal results that we want. A series of moral and social objections could be launched against this idea."

Feminism, Individualism, and Liberalism

Richard Stith returned to the earlier discussion of communitarianism as a major alternative to individualism. "Why are there no feminist communitarians?" he asked Elshtain. "Is feminism inherently hyper-individualistic? I remember there was a time in the late sixties when many of us, myself included, were going the other way. To talk of something as being personal or of private ownership was counter-revolutionary, was wrong. And when feminism emerged at that time, right from the beginning it seemed—because of the abortion issue—to speak about individual, privatistic rights. Then I thought that feminism was really counter-revolutionary because it was going back to privatism. One would have thought radical women in the 1960s would have taken the traditional Marxist route of freeing their gender—the proletarians—as a group from oppression and becoming the dictatorship. But instead the solution was 'How can we help individual proletarians become white-collar workers?' Why did it have to take that turn?"

Elshtain responded: "The reason is that if you're working from a hypothesis of universal female oppression in every pre-existing social form, in every historic ethos, then you're bound to be not just suspicious of but hostile to communities, because those are the places where women got oppressed all these years. The 'motherhood' image—with the community of the family—was itself seen to be oppressive. The only acceptable image was 'sisterhood,' but that is a very watery thing. It was evoked to keep people in line when there might be ideological defection. The only acceptable group was a wholly voluntaristic group, coming together for consciousness-raising. But those groups didn't last long. There are some feminists—they're mostly Marxists—who argue that at some future point, when we have cleansed the social world of the oppression of women, women will be able to be communitarians. But they don't use that language; it's more like the 'new socialist order.' The whole language of community seems enormously suspect—although among some feminist theologians there is talk about community, about new spiritual communities."

"How is 'liberal' different from 'ultra-liberal' in terms of providing a different account of the self?" Stanley Hauerwas asked Elshtain. "And how does that help you discriminate between legitimate and illegitimate interventions? I assume, since you want to continue to identify in some sense with the liberal project, that you have a different psychology of the self than the ultra-liberal does, and I want to know how that will help you provide discriminating ways of knowing why certain kinds of interventions might be legitimate, and other kinds might not be. For example, why might artificial insemination by a woman's husband be legitimate, but artificial insemination by a donor not be?"

"I'm not sure how much help it would be to sort out particulars of those kinds of things," Elshtain answered. "As you know, liberalism is many different sorts of things, which is why liberals have all kinds of arguments with one another. My point here was, in one sense, a practical one. I don't think we could jettison the liberal project if we tried. It goes back to yesterday's question: Are there alternative ways to sustain and protect the accomplishments of liberal society that we all acknowledge? Is there some resituated, transformed understanding of rights?"

Fred Sommers, returning to the question of handling the problem of technology, entered the discussion. "It seems clear that we're committed to some amount of technology. Since we're not Christian Scientists, we agree that we sometimes have to use technology in an interventionist way. The question is, How far do we go or how far can we go, and where do constraints come from? And, of course, one way to produce constraints is simply to adopt a theological position—from that position you work out the theology, you work out some of the constraints.

"Another way to develop constraints is to take the analogy

of ecology seriously enough. It's interesting to see how liberals handle ecology. They have a reverence for tradition provided it isn't 'too much' tradition. If something is 'natural,' there's a certain reverence for not disturbing that, and a piety about it. But the family, for instance, which has been around for a long time, is up for grabs. Liberals can play around with that. But they don't believe in playing around with systems that come up 'naturally,' as they put it. Why do feminists treat the established culture, the traditions of culture, the institutions of culture as if they were not 'natural' in some way and have no normative force for us? One of the reasons we're in crisis now is that for many reasons traditions have eroded. This hasn't happened because of the feminists or the technologists. It has to do with other forces and reasons. Many today have no respect for 'what's there,' and yet there are all these new possibilities. The work that has to be done can't be done just intellectually, with a priori reasoning. It has to be done politically, by trying to get public policy that reinforces tradition, reinforces the family, reinforces the kind of education we want to see. If we do create a population with a greater respect for tradition, then we will have some chance of imposing some constraints on technological innovations that alarm us and should be alarming us."

"I'd like to know the policies you have in mind," said Brigitte Berger, "because each one that has been tried has failed."

"I'm not a public person at all," Sommers answered, "but let's start with tax policies. We should work for tax policies that reinforce familial arrangements and work against the disintegration of families."

Neuhaus came to Sommers' aid: "In this morning's *Wall Street Journal*, there's a list of six or seven policy proposals that would do precisely what you suggest—create economic incentives for stable families."

Neuhaus went on to summarize Sommers' remarks in terms of the question to be addressed in the next session on James Burtchaell's paper: "There's the whole question of a 'human ecology'—that includes the institutions which have seemed to be amenable to the flourishing of the human condition and how the understanding that undergirds them can be advanced. The other matter—which I hope other people will come back to—is the issue of how far we are prepared to go with regard to technological interventions. And I wonder if that's the right way of putting it. Perhaps the question is not how far but to what ends we are willing to intervene. That is a much more tricky thing."

The Family

At this point Stephen Post spoke up. "I like the conclusion of your paper," he said to Elshtain, "because it seems to me you're saying that feminism is becoming increasingly pro-family. That change is pure gain. What I want to know is, How pervasive is that? Betty Friedan was originally anti-family; now she's becoming pro-family. There are a lot of feminists who are moving in that direction, working to retrieve the notion of an embodied, familial kind of philosophy of the self vis-à-vis liberalism."

Elshtain agreed that feminists are becoming more pro-family. "But it's a tricky enterprise," she explained, "because most feminists don't want to endorse a norm of the 'ideal' family. They see that as oppressive to alternative family arrangements. There is complexity in their being pro-family, and they want to leave the definition of the family totally open.

"The family is a deep problem for liberalism," Elshtain continued, "because it seems to represent all the sorts of things that liberalism helps us to get over—for example, the principle of authority in the family. It's not chosen; we don't elect our parents. So how is this form of authority justified? The ties of blood are not contractual in an ideal sense, and all the rest. So the family has always presented great difficulty for liberal political philosophy."

Gilbert Meilaender of the religion department of Oberlin College asked Elshtain a related question. "You say in your paper that feminism may be forced to turn back to reconsider the family. Suppose, to the degree that this happens, it will be the family conceived in much more contractual kinds of terms, in terms of role relationships and responsibilities and that sort of thing. So there will be a return to this natural community of the family, but it will be reconceived in liberal, maybe hyper-liberal terms. Will that be a net gain as far as you're concerned?"

Elshtain answered that the heyday of the demand of earlier feminism for detailed contracts seems to have passed. "There seems to be less emphasis on that and more on pop psychology, on what people feel good about. The contractual emphasis is incredibly privatized and totally disarticulated from any sense of intergenerational continuity and any sense of the family as something that radiates outward. The norm of the family that I unabashedly endorse is one that respects tradition, intergenerational continuity, with the children growing up to be decent, responsible citizens."

Sidney Callahan offered this observation: "We're in a postmodern moment where we're rethinking science, rethinking reason, rethinking nature—a lot of things that we've accepted until now. I think we have to accept that human beings are natural technocrats, natural users of technology; this has been true from the first time a person picked up a stick. This is what we were made to be in the image of God; we were made to work on the world and to accept technology. Our problem is how to humanize it, how not to let it get out of hand."

"You mean working on the world and working on ourselves?" Neuhaus asked.

"I think so. Transformation. So I think I am a 2 Corinthians Gnostic and therefore believe in the universal part of the message which sanctions the Enlightenment project, the liberal project, and feminism. Let me put in a plug here for feminism. We're acting as though the crazies in feminism are keeping us from realizing how important this movement has been. My own experience of twentieth-century women is that they have not had an adequate sense of self, that they have been too much centered in the bodily sense of self. I can understand why the feminists went in the direction they did, because they developed a sense of all the other things that have to do with selfhood. I think this sense of selfhood comes out of Christianity: 'In Christ there is neither male nor female, neither slave nor free.' So I see feminism as essentially based in Christianity. In the early Christian tradition the first feminists were orders of sisters who had the idea that it wasn't necessary to be in a family. Families can be very oppressive institutions, and we have to remember that when we're trying to restore the tradition. Although I recognize all these other problems, I see the maternal strain in feminism as being very creative. And I think the issue is to keep trying to have feminism develop rather than saying, 'Scratch that.' By the way, there's a very strong feminist pro-life group gradually gaining adherents. So isn't feminism redeemable? Couldn't it be a creative project?"

"So you envision a dramatically different feminism?" Neuhaus asked.

"Let's try a Marian view of feminism—a whole, maternal, much more humanitarian view, which I think has always been there," Callahan replied.

Brigitte Berger commented on the trend in feminism of returning to the family. "That's all to the good, but I don't think it's essentially changing the situation. Here I would differ with Professor Callahan. It's important to understand that feminism is simply one manifestation of a much larger process in the Western world—what I call the process of modernization, which is carried on by a very distinctive class. Let's go back to the notion of the new feminists, this new discovery of motherhood. Yes indeed, they have now discovered motherhood, but it's not motherhood in the traditional sense—it's liberated motherhood. A woman *chooses* whether or not she wants to use her reproductive capacities. We noted that no-fault divorce has been a failure, but what is the solution? More government intervention—more money, more control, more entitlements, more child-care allowances, more of the same. Under the guise of a recovery of motherhood, we have the same forces marching on. The class force is what I would call the 'new class,' which understands life in atomistic components that can be put together in various forms."

Elshtain addressed Callahan's point, noting that the kind of feminism Callahan was talking about has a long-standing historical presence in the United States. And she agreed with Berger's comments about feminism as part of a wider process of modernization.

Christopher Lasch directed attention to an element missing from the discussion. "If the problem is limits and the refusal of modern liberal culture to recognize any, it's not clear to me how this discussion of technological interventions can be conducted without any reference to capitalism. After all, it's the vision of material abundance founded on the absolutely unprecedented productiveness of the capitalist economy that underwrites all this and makes the vision of a life without limits, without any constraints, so plausible. I don't see how you can attack the plausibility of that conception without also indicting its association with the whole history of modern capitalism in the West, which is by its very nature dynamic, expansive, uncontrollable. And of course one could make all sorts of more specific connections between the history of capitalism and the history of modern technology. They simply can't be divorced."

"Isn't the feminist thing an instance of that drive toward life without limits?" Neuhaus asked.

"Absolutely," Lasch answered. "What we seem to have here at times is a very broad-gauged criticism of liberalism that has to be extended to the market economy. In our current political discourse, liberalism has a very specific meaning, and it's assumed that if you criticize liberalism, you must be some kind of 'conservative' according to the current meaning of that term. That is, you must in some way be identified with the New Right. Well, the New Right has no quarrel with limitlessness, with uncontrollable technological development. It has been much more interested in defense of the free market than in the things we're talking about."

"One quick point in closing," Berger added. "I always come back to the religious question. Human beings, not understanding what their nature is, have to turn to something other than themselves. Traditionally this has always meant turning to a higher order, to God. That is how we have always understood human nature," she said. "However, once we lose this imbeddedness of human nature in the religious tradition, we have to turn elsewhere. When we are no longer responsible to God, we can turn to experimenting with ourselves."

THE CHILD AS CHATTEL: REFLECTIONS ON A VATICAN DOCUMENT

In introducing the final session of the first conference on "Guaranteeing the Good Life," moderator Neuhaus reminded the conferees of a subject that Francis Canavan had raised in the previous session, the "repugnance point" that might trigger fundamental reconsiderations—philosophical, spiritual, and so on. There are times when a repugnance point is reached on major issues, he noted. "It happens when enough things go wrong, and enough people are alarmed or disgusted by what has gone wrong, that they say there must be some wrong ideas behind the issue.

"A good example would be welfare policy in the last twenty-five years. It's a fact that in the Bedford-Stuyvesant section of Brooklyn last year, 83 percent of all births occurred outside the context of marriage. The implications of that are widely recognized and have occasioned wide revulsion and reconsideration of welfare. There is indeed a new consensus today, across the political and ideological spectrum, that the problem facing welfare is not simply providing more generous benefits or fiddling with this policy or that, but the problem of patterns of dependency. Here's an instance of a repugnance point being reached and forcing major reconsideration. Of course, the price paid by the bottom third of black, urban North America has been tragic. And obviously a price is being paid in the process of reaching a repugnance point with respect to the issues that our conference is now addressing. What might happen as the aging and the handicapped become increasingly aware of the implications of the return of eugenics for their well-being?"

Moving on to James Burtchaell's paper, Neuhaus noted that the ecology motif had entered the discussion earlier. "There's a marvelous continuity here. This morning Fred Sommers got us started on the question of nature and ecology and how that enters in. Then Brigitte Berger and others picked up on the distinctions that need to be made in how we use the term 'nature' and the different kinds of appeals to nature. All of that leads us very directly into Jim Burtchaell's presentation."

Burtchaell, by way of background, related the story of how he came to write his paper. He recounted a pilgrimage that led him, after years of reflection on the church's position on artificial conception, to the conclusion that the church is a community with accumulated wisdom. The church passes on to us not commands but wisdom. That wisdom, however, comes in an imperative voice.

"The argument I'm making is that there is traffic between what we do and what we are," Burtchaell said. "The church has an inveterate experience; therefore, for example, the church is a great source for finding out what happens to us when we steal, when we kill. The papacy has been in a position to draw prophetically on the pastoral insights of centuries." He was critical of certain specifics in the Vatican document that he treated in his paper. Regarding contraception, he said, the main point in the church's position should be not the distinction between artificial and non-artificial means of birth control but the entire story of sexual encounter between a couple within marriage, understood in terms of welcoming the stranger who is to come.

Burtchaell continued, "Neither love, as Joseph Fletcher would claim, nor choice, as the people we've been talking about these past two days would say—neither love nor choice makes acts good or life-giving. And the experience of the church—which sees Jesus crucified as God's most poignant disclosure of his stubborn, persistent cherishing of us—sees that there are certain ways we need to behave (in this case, reproductively) in the bearing and cherishing of children. Otherwise we will not flourish; we will wither."

The "accumulated wisdom" interpretation of the church's contribution to moral discourse on the issues before the conference, together with the question of whether its teachings are directed to the community itself or to the world at large, became a major issue in the early part of the afternoon's discussion. A second major theme became Burtchaell's image of the family welcoming children as a basis for moral decision-making growing out of the church's accumulated wisdom.

The Role and Contribution of the Church

Stanley Hauerwas opened the discussion. "I want to speak in favor of the Vatican for a minute, as a way of trying to sharpen the issue. The reason the Vatican wanted to move to a naturalistic account of the *telos* of the generative act was that it wanted to make the moral recommendation apply to everyone. In the encyclicals the pope doesn't want to write only to Catholics; he wants to write to everyone. The Vatican recognizes that if it has to have special theological presuppositions to sustain the view of marriage, it can witness to what that kind of marriage means but it cannot impose it. The Vatican wants to impose, and therefore it uses a natural-law argument. If Father Burtchaell wants to make these arguments a matter of wisdom rather than a matter of necessity, in terms of those kinds of appeals to the natural, then the Vatican is going to pin him on this.

"For example, Father Burtchaell, in your paper you say you're open to AIH [artificial insemination by husband]. But if you're open to AIH, why aren't you open to AID [artificial insemination by donor]? Why isn't that a furthering of the intention of the married couple's will to have children in that way? If you're open to AID, why aren't you open to in vitro fertilization? You can argue consequentially and say that these procedures just open up things that we'd best keep closed. Still, they're matters of wisdom, and you can say, 'I can surround this procedure with proper safeguards in a way that it won't be open to misuse and so on.' What the Vatican does is argue on the straight physicality of the act of sexual intercourse. What you do is put the whole argument in terms of the wisdom of the church about how sexual intercourse serves the ends of marriage, not the ends of nature. You want to say that lifelong marital fidelity is natural, given that we're God's creatures. But that's a point of witness, not a point of necessity. The Vatican understands what this means for public-policy issues, and how difficult it then becomes to say 'no' to something like artificial insemination by donors."

Burtchaell replied, "Whatever the Vatican's concern about legislation on reproductive procedures around the world, it has a prior obligation to share with its own members what the implications of their own faith, and their own faith-illumined experience, might be with regard to reproduction. For that reason, while Professor Hauerwas may be correct about why the Vatican was obliged to use biological teleology, there is still great loss in its having done that. Hauerwas sees a fairly strict dissociation between what I was calling 'moral wisdom' and what he calls 'necessity.' I'm uncomfortable with that, because the type of wisdom I'm talking about is not wisdom about what is true only for members of the church but wisdom about what is true for all those who engage in sexuality. It is the church's own perspective on that, but it is a claim that all people who engage in sexual partnership need, for their own benefit, to follow the call of that experience into monogamous fidelity. Whenever you amputate the teleology of the sexual partnership and consider only its biological aspects, you do violence to it. The church's message has to be that sexuality requires monogamous fidelity, and that the sexual partnership must be a commitment not only to that other party but also to the children who will be theirs."

"The question I have for Father Burtchaell," Jean Elshtain said, "is a political one. People outside the church are automatically put off by a perception of the church as authoritarian, as ordering people around. How is it possible to alter that particular perspective? Is there something that can be done about that?"

Burtchaell replied that the church conveys heavy moral conviction and wisdom in a voice of authority. "Real, effective authority," he continued, "must explain what is at stake. That's why testimony is so important. The teacher in the community must represent enough of the testimony, persuasively enough, to let people learn. I like to call authority a wisdom, but it is an authoritative wisdom. It says, 'You stand to live or die, spiritually, on the strength of what I have to tell you.' What makes it sound totalitarian is someone's saying, 'Trust me' without explaining, or 'I'll punish you if you ignore what I have to say.' That is a total failure of nerve, which has so often gone on, and so often has destroyed the ability of anyone in the church to be instructed to his or her benefit."

The Family Welcoming Children as a Basis for Moral Reflection

Jonathan Imber brought in the family image when addressing Burtchaell about his paper. "I found quite inspiring the way you developed your own conceptualization of the Fifth Commandment: that to honor your father and mother works both ways— toward children and toward parents. In that regard it seems to me that all the new reproductive technologies serve to confuse us. The more we make it clear that there is that kind of confusion, the more people will be paying attention to the kinds of outcomes that occur in all of the experiments that are going on now. I don't know that we're going to stop them. But I do think that by directing our attention out of these traditions to what may be the untoward, unexpected, even horrific consequences for individuals as a result of these technologies, we will pay attention in a different way."

Robert Destro picked up the point with the notion of "welcoming children." He mentioned Burtchaell's reference to not having had the experience. "This is something that's rather fresh in my recollection, because my wife and I were a fairly typical 'Yuppie' couple who had to go through some of this, since we're not able to have kids naturally. I don't want this to sound like an

'Alcoholics Anonymous lead' or anything, but it pertains to Stan's question about why, if you accept AIH, you don't accept AID? The anterior question is 'What's so wrong with AIH if it fits into the general notion of the husband-and-wife relationship over the long term?' And I think it connects with a number of things in my own experience. One of the first things my wife and I were instructed to do was to attend a meeting of an organization called Resolve, which is like Alcoholics Anonymous. You admit to everybody,'I'm infertile'; it's just like saying 'I'm an alcoholic' at an AA meeting. At our first meeting my wife and I reached the point of repugnance.

"I think this experience relates to the notion of welcoming children and the nature of the relationship between husband and wife. Welcoming children is different from questing for them. In questing for them, you go through some of the tests, and you feel like a piece of meat. You see that the primary relationship isn't between the husband and wife at all; it's between the wife and the doctor. When my wife and I went to these meetings and listened to other people tell their stories, there was something about going through the process that made both of us uncomfortable. My wife was probably more uncomfortable than I, because I'm much more into the whole notion of technology. But the point is that somebody gets left out of the process: the husband gets lost. With a donor that's absolutely true. With in vitro fertilization both the husband and the wife get lost. The wife simply becomes the object. But in AIH the husband gets lost too, and it's not because of the sperm in the tube. It's because the relationship itself is sundered, and the husband isn't a part of it in the same way he is in the natural way. And now, as an adoptive parent, I can say that technologically assisted procreation is not the same as welcoming an adopted child into the family, because in this situation both husband and wife welcome the child; both are equal partners in commitment."

"There are third parties in adoption," Hauerwas noted.

"Well, there are," Destro responded, "but there's a minister when you get married, too. They are witnesses to the adoption rather than parties to it."

"But if in vitro fertilization were made much easier, people wouldn't have to go through these dehumanizing processes," said Brigitte Berger. "It would be done in just a flash, and there the child would be. Do you think that would be all right?"

"I don't think it can ever be made that easy," Destro answered. "My wife and I recoiled from the process because it's not—I hate to use the term—'natural.' There's something about that wisdom, and I think Father Burtchaell is on to something when he talks about the church's collective wisdom. The wisdom arises from the nature of what's going on, the nature of this little community you have."

Burtchaell said to Destro, "Your testimony is the best example we've had so far of what I would describe as the wisdom discourse within the church. Now your testimony would have to be confronted by that of other people who probably were helped by much better doctors. Unfortunately, this particular type of medicine is largely in the hands of research medics, whose bedside manner would be roughly that of Adolph Eichmann.

"I think the church's wisdom discourse would be compared to the highly valuable conflicts about breast-feeding and bottle-feeding today, and about the Yuppie mother who slaps the kid into day care on day four. We will come to our wisdom only in this way. But it is a wisdom about necessary items, not about optional ideals. It is a wisdom about our inner survival needs. It may end up as something on which the house divides; it may end up as something that involves a lot of cautions."

Neuhaus interrupted. "Jim, you've lost me. You said that Destro's testimony must be linked with other people's testimony, and you said that we have a similar disagreement about breast-feeding and bottle-feeding. Then why isn't this wisdom simply a matter of what every couple finds fitting for them? Does it have any kind of universal claim or any kind of necessity? Why isn't it just 'Do your own thing'?"

"The church is a communion," Burtchaell replied. "That is, the church is a dining society, and at this dinner we argue. Right now the issue of reproductive technology and the family is an unresolved and highly debated issue in the church. We're all here at great cost to think about it. And we stand to gain by continued discourse. As that discourse goes on in the community of belief, it swells to some sort of resolution, generally. We can't control when that will happen. We might be faced by decisions that have to be made, but it's very painful to make those decisions when there isn't good resolve. I think it is the task of those who preside in the church to orchestrate this encounter, this discourse, this sometimes very conflictive set of testimonies."

"On, for example, the kind of distinction between AIH and AID?" Hauerwas interjected.

"Yes and no," Burtchaell answered. "I see those good families who have turned to AIH and have been assisted by the right kind of physicians generally perceiving the procedure as having enhanced their wedlock and their family and not intruded upon it. AIH has been intervention but not intrusion. But I see the church, from much longer consideration, saying that the intervention of another man's sperm is no different from the intervention of another man—it's adulterous. Now, the church's judgment on adultery is based on the experience of what happens to people when they sleep around. But I find no sufficient reason in the community of faith, no reason that has gained any foothold in our consciences, for arguing that adultery is, after all, 'just kind of helpful once in a while.' "

"Jim, I really do want to ask a few questions," Neuhaus said. "We're flipping back and forth between this matter of the wisdom of the community and something in the Vatican document that, as you yourself point out, is presumably applicable in a more public way, even to the degree of public law. You say you have these conflicting testimonies, and at some point the community of faith says, 'Aha!' What does this experience in the community of faith mean to the larger society? What bearing does it have on the question of what one is publicly going to claim as being applicable to the ordering of the lives of people who are not part of that community of faith? It's your position on that cluster of questions which I'm very puzzled about."

"I see at least two issues here," Burtchaell replied. "The community of faith must argue according to its own understanding of this, and it must receive testimony. Destro said that, and by the way in which he said it he left himself open to other people who might have contributions to make. That, in microcosm, is what the community of faith must do. Eventually enough theologians will be brought in for the concept to be formulated in a way that will endure."

At this point Hadley Arkes entered the discussion: "Would you say that the church would be drawing lessons of experience, formulating its rules, and that certain of these lessons must be gross? On the matter of adultery, we come out with a very gross and emphatic rule without qualification, because that's our reading of our experience thus far. But could that change in another period, when sensibilities change? Or are you actually saying that there could be some universal rules drawn as safe predictions about an enduring human nature that would give us confidence that these rules would endure?"

Burtchaell responded, "I think the community of faith has some convictions of wisdom that it considers to be stable precisely because they are comments about how we are. On the other hand, the community realizes its own fallibility and capacity for growth, and it remembers new lessons learned—not that the way we are has changed, but that we have come to know it better— and therefore has made some important changes. It's a comment about the changeability not of how we are but of our grasp of how we are."

"Regarding Bob Destro's issue," Arkes continued, "do you consider that question still fluid and unsettled?"

"Yes," Burtchaell responded.

Destro picked up on this reference to his own experience. "I've talked to a lot of people about this procedure, and there certainly can be aberrations. The people who have gone through it and feel fulfilled say it was worth the violation to now have the result. But nobody denies the violation. And that's what I thought Father Burtchaell was talking about—about the church's wisdom."

"No, no," Burtchaell replied. "I know some people who don't fit that description. They don't call it a violation. A cesarean is a violation, too, but how a woman feels about it depends on her sense afterwards of how the event fits in with things."

Is Technology Morally Neutral?

Next Barry Schwartz returned to a point he had raised in the morning session with reference to surrogacy—whether the moral judgment might have more to do with the intent surrounding the act than with the act itself. He had the same feeling about much that was raised in the Burtchaell paper, he said. With regard to salting the Massachusetts turnpike or using up the water in the aquifer, the causal relationship between the acts and the harmful consequences is unambiguous. But more dubious is the paper's attribution of teenage pregnancy, sexual promiscuity, venereal disease, abortion, marital collapse, child abuse and abandonment, and so on to "the venture of technology." That might be true, he continued. "But it's very hard for me to see that the relation between reproductive technology and these evils is obvious in the way that the relation between soil erosion and salting the highways is. The technology is plausibly neutral with respect to these evils on which we all agree. And what we need to ask is whether the technology could be reconstituted or contextualized in such a way that it achieves good ends without the accompanying evils. It seems that there are certainly problems in the case of in vitro fertilization, AIH, and so forth. One problem that many people experience, I suspect, is that these are regarded as technical matters to be achieved by technicians, and these people get counseling that isn't so much spiritual as it is psychological. It seems to me that it isn't necessarily a violation to go through this kind of AIH if in fact the couple is given the appropriate spiritual support—support that affirms the bond between wife and husband and their joint commitment to their offspring. Now maybe it's difficult to arrange that—but no one has even tried because people don't see spiritual support as the thing that's missing. What they see as missing is the ability to produce a fertilized egg.

"Now spiritual support is not in the logic of the technology at all. It's in the way people understand what technology has to achieve. And this is what has to change. They think all technology has to achieve is fertilization. In fact, it has to achieve fertilization with the appropriate kind of commitment and attachment and bonding that go along with fertilization accomplished in the old-fashioned way."

Neuhaus pressed Schwartz on this point. If the problem is infertility, understood as "how to get this sperm to this egg under suitable circumstances," he said, "there is an inescapably technical answer. Are you saying the couple shouldn't define the problem simply as one to which there is a technical answer but should define the problem more broadly, so that technology becomes part of an answer more coherently related to their entire relationship?"

Schwartz responded affirmatively. "When you solve a narrow technical problem, it turns out that that is only part of the problem. Fertilizing an egg is not all that couples with fertility problems want and need from the reproductive act. So other things in addition to the technique for fertilization need to be supplied."

Schwartz went on to suggest that the modern expectation that both husband and wife enjoy sex has placed an additional strain on monogamous marriage, leading to a need for instruction in techniques that allow both partners to get mutual satisfaction out of sexual activity. "The Dr. Ruths of this world," he commented, "could be construed as helping to solve a moral problem by instructing people in the technology of sex so that it's possible for them to maintain a bond to eternity. Perhaps the quest for this satisfaction should be conducted by the church. The church is the institution with the most powerful stake in the lifelong sanctity of marriage."

"In any case," said Neuhaus, "you're saying we shouldn't blame technology for all the problems that may surround its use."

At this point Richard Stith asked whether technology is intrinsically evil. "I really have the feeling that pride and liberalism and all kinds of terrible things are written right into science and technology, and that since the time of Bacon science and technology have been informed by atheism. They are microcosms, and the macrocosm is atheism. They are an attempt to dominate and control the world completely for our own self-interest, the supporting belief being that there are no limits to that control."

Now Sidney Callahan picked up on Schwartz's reference to Dr. Ruth, a former colleague. "I do believe in sex education. I see technology as the imitation of God's creation of the world, so I'm very much in the pro-technology camp. If we really understand something, we can intervene in a much better way. I guess my image of this is so-called natural birth, which is not at all 'natural,' because women who choose it have to go through a very long training process in which they obtain knowledge, practice breathing control, and so on. But it's a wonderful example of how knowledge and intervention can enhance life and even take away pain. So I see sex education and this effort to improve things as good. Technology can be used in such a way that it goes with the world and with 'nature.'

"In the end," she said, "I come out against surrogate motherhood and in favor of AIH and contraception, because the whole is greater than a single aspect. As far as AID is concerned, I'm against third parties. But, generally speaking, I think technology can be used for good, and that with it we're going to try to do good things for the world."

"So you come down on the side of technology being neutral?" asked Neuhaus.

"No, I don't think all technology is neutral. After all, certain kinds of technology can be used to kill people," noted Callahan.

Jean Elshtain countered that technology can't possibly be neutral. "That doesn't necessarily mean that it's evil. But it always emerges out of a particular historical matrix, it requires a political economy that can command resources, it's entangled with the instrumentalist worldview, and it helps to generate demand. Technologists find ways to do certain things, and then they retail them, and that in turn generates demand. They say, 'Look, people need this.' A structure of needs emerges out of this complicated socio-political, technologized world. When infertility got medicalized and a solution to it seemed within reach, we thought we could correct the shortcomings of nature. That's when demand began to arise, and technologists said, 'People need this.'

Christopher Lasch agreed. "It's a very serious mistake to think of technology as neutral. Every society develops a certain kind of technology that reflects its beliefs, its values, and its ideologies, which cannot exist without an elaborate framework of institutions and laws. One of the consequences of believing that technology is neutral is debating every intervention on its merits and deciding on a case-by-case basis whether the use of technology is good or bad. That won't work, and the awareness that it won't work is one of the few hopeful signs in our society. These choices are cumulative, often irreversible in their effects. It's impossible to predict with any assurance where all this will lead.

James Burtchaell wrapped up the subject. "I resonate with the view that it isn't technology itself which is unacceptably intrusive. But I wouldn't want to say that every adjunct of technology can be redeemed if we can just find the right manner. I don't quite share Richard Stith's view that all science and technology are satanic. I thought of the development of certain new varieties of wheat and strains of rice that are feeding the world. This is as disciplined an intervention into nature—in order to allow the earth to produce—as some of the things Sidney was talking about. The community learns from experience how technology can enhance—and suppress—what we need, and often technology comes in with dominion rather than with service. Technology, science, and engineering can be helpful, but sometimes the only way to understand an act is to understand who is doing it. Where procreative technology is concerned, the greatest caution that we had, which caused many of us to withhold assent from the entire program of *Humanae Vitae*, was our close knowledge of the many wonderfully generous couples for whom contraception had meant neither a violation of their union nor a surrender to the zeitgeist that concerned the pope in the first place."

As the final session of the first part of the conference drew to a close, Neuhaus suggested focusing the remaining discussion on certain points: "The issue that brings us together is our uneasiness about ideas on and actions and trends toward the human control of the human condition, the elimination of contingencies from the human condition, which is what eugenics is. This afternoon Father Burtchaell has turned our attention to the truth that in nature there is a givenness to be respected or to be violated only at enormous peril. And that is built on what was said this morning. I was particularly struck by Fred Sommers' suggestion that what has happened in the past twenty years in terms of the environmental consciousness, the ecology movement, and so on is very hopeful, and now the hope is that people will begin to see something like human ecology, the cumulative experience of what makes for the flourishing of human beings. Brigitte Berger was not enthusiastic about this idea; she said that to look to an appeal to nature is to abdicate what is really the only hope we have, which is something that transcends nature, which holds us accountable as creatures who will be called to judgment. This is something we might spend some time on—the usefulness and the dangers of the appeal to nature.

"Then we might examine Frank Canavan's question: What will force reconsideration? What will force people back to the kinds of questions that Kit Lasch and Hadley Arkes were getting at—namely, *the* political question of how we ought to order our life together; what are the principles, the ideas, the visions by which we order our life together?

"Let's concentrate on these two issues: the usefulness and

the dangers of the appeal to nature; and what we expect to be points of argument, experience, and crisis that in the foreseeable future will force reconsideration."

The Appeal to Nature

Jean Elshtain questioned the appeal to nature. "Appeals to nature and natural law carry much less weight with us than they might have at one point. Unless it's 'crunchy granola' and part of a 'natural lifestyle,' nature becomes that which we want to control, alter, and transform to suit the needs we have generated."

Francis Canavan suggested that technology is part of being human. "If we set it against nature, we're headed for disaster. We can't simply ask whether technology in itself is good or bad. We have to ask about its relationship to nature, and its particular uses will be judged by reference to some natural norm. Those of us who believe that nature is the creation of a personal God have a vastly different attitude toward nature than those who don't believe so. For those who are atheists or pantheists, it becomes much more questionable why nature should be the norm."

Stephen Post asked if theism is necessary for this agenda. Straightforward, consequentialist kinds of arguments are possible. "Artificial insemination by donor is different from artificial insemination by husband," claimed Post, "because it has a negative impact on the identity of the child. And identity is crucial to our sense of integrity and so forth. I think one could appeal to nature without necessarily appealing to the Author of nature, even though I think a lot of us believe in that Author."

"Certainly it's possible to appeal to nature without appealing to God," Canavan replied. "We do that all the time. But it doesn't mean quite the same thing."

Next Fred Sommers entered the discussion. Regarding the appeal to nature and the necessity for the theistic dimension, he said, there are two aspects: "The negative question is what happens when we fool around too much with social engineering and mechanical intervention." In the natural sciences, he pointed out, the consequences of particular acts are predictable. Not so with social interventions. "I don't think we need to appeal to God to know about the monstrous consequences that have occurred because of thoughtless interventions. Take the whole approach of the social engineers in, say, Marxist interventions, and their disregard for the individual. We don't need to appeal to God to know that such interventions are terrible."

Brigitte Berger, noting that her paper (to be presented at the second meeting of the conference) would address some of the issues being raised, emphasized the difference between the natural scientists and technology, and the social scientists and new professions who have contrived a body of knowledge for internal reference. "For that reason," she said, "we come back to the question of religion. The only thing that keeps one honest is some other standard, a transcendent Being."

"The conscience might do that," said Neuhaus.

"But where does one's conscience come from?" Berger asked.

Stanley Hauerwas returned to the question of nature: "I get very nervous speaking about nature qua nature. It's a term without context. I suspect that the word 'nature' only works with a Christian theology, as a way to distinguish between the 'supernatural' and the 'natural,' which is not the way we're using the term at all. This is the way we're using it: if we say that infertility is unnatural, that means it should be subject to human intervention, so we medicalize it. These descriptions of infertility, abortion, and infanticide are not rooted in something called 'nature.' They are rooted in something like a community's sense of moral wisdom that has been gained through the experience of a people that has been formed virtuously, that has discovered these things."

"Tell me, Stan, how you make that argument to someone who says either 'I don't respect what the community tells us' or 'I belong to a community that has a different wisdom'?" Neuhaus asked. "How do you make that argument publicly?"

"There's no other way to do it," he replied. "There is no 'nature' someone can point to and say, 'My views are anchored in the way things are, and yours aren't.' If a Muslim argues about marriage, he's going to say polygamy is anchored in the way things are. There's no way to resolve kinds of those disputes. The liberal assumption that there is some kind of inherent rationality which is separate from the moral training of good communities is wrong. For me, in the end, appeals to nature won't cut it. Much more determinative are the kinds of moral convictions that have trained the community to think this rather than that."

Canavan asked, "To what do you appeal for your standard of what is a good community, or a moral community, that is ultimately something objective and not man-made?"

Hauerwas answered, "Of course I'm a Christian. I didn't create the beliefs of the convictional community that I think offers a way of life God has called all people to. It calls me to activity, both intellectual and moral, to try to articulate, for myself and others, what it means to flourish."

"So then, like the Muslim, you believe that your moral judgment is in accord with the way things really are for everybody, although not everybody perceives this," Neuhaus said. "And the way things really are, ultimately, is what some people mean by nature."

"Yes," Hauerwas responded, "but that's within that ongoing community of discourse."

Gilbert Meilaender spoke up, noting that appeals to nature make more sense to him than to Hauerwas. "Appeals to nature are trying to say that there are limits, natural and historical restraints, that are supposed to be respected. Human life, human society, is not to be thought of primarily as something we fashion. The question is whether there are some things which in our freedom we could do and might wish to do but which we nevertheless ought not to do because they would be destructive of our nature. Doing those things would somehow be going against the grain of the way things really are. It's not just a question of bad consequences. To do those things would be ontologically destructive, even if useful in certain ways. There's a difference between surviving and flourishing. Sometimes an exercise of our freedom would be destructive of the nature of the human person. That's what appeals to nature are trying to get at."

"Why are statements about the way things are statements about nature?" Christopher Lasch asked. "They're not. They're statements about nature, to be sure, but they're also statements about man's life, which is in nature but also not in nature."

Following a discussion about what was meant by 'nature,' Barry Schwartz re-entered the discussion. "Appealing to naturalness is fraught with difficulties," he asserted. "The people who do it best these days are biologists, and their appeal to naturalness is an appeal to continuity between people and chimpanzees, ants and cockroaches. Whatever one might want to say about cockroaches, moral discernment is not one of their characteristics." Human evolution, he went on to say, has taken place within cultures. The natural state of man, therefore, is to be dependent upon cultural institutions. "The problem with that is that it doesn't resolve any questions about what's good or bad, because there's so much diversity among cultural institutions. Appealing to naturalness doesn't solve those questions. It just pushes them to a different place."

What Will Force Reconsideration?

Addressing Neuhaus's second question about what will force reconsideration of technological interventions, Schwartz went on to note the relationship between moral commitments and industrial, market-oriented societies. "What people won't stand," he said, "is the economic collapse that will follow as a consequence of the commitment to individualism that underwrites the technology that is ostensibly the purpose of our gathering. It is not so much abortion or in vitro fertilization or infanticide that will be the point at which people will say, 'No more.' That point will come when they can no longer satisfy the desires they confuse with needs, in part because of the moral decay that we're talking about."

Neuhaus summarized, "So your point is that reconsideration is going to occur not because people at one point or another say, 'That's an abomination; that's morally wrong; that's murder,' but rather because they will be frustrated in their attempts to satisfy what they perceive to be their needs."

Sidney Callahan spoke to the same point of what will force reconsideration. "It's possible to learn that overcontrol produces disaster," she said. "I think here of Erik Erikson, who said that the only way to control something absolutely is to kill it. What will force reconsideration is if the things most important to us are killed. What are those things? Our human relationships—our friendships, our family relationships, our emotional commitments and bonds, particularly in marriage. So if we see that abortion is producing post-abortion syndrome, or that people are not able to love, or that adolescents are committing suicide—that there is this breakdown in our ability to have bonds . . ."

"But people are seeing that now," countered Neuhaus. "How much more has to be seen?"

Callahan pointed to movements like Alcoholics Anonymous, in which individuals achieve control by letting go. What really matters to people is family life and emotional relationships. She suggested that at some point the deterioration will reach a critical juncture which will force reconsideration.

As the first conference drew to a close, James Burtchaell made a closing statement. "The believing community," he said, "tends to be like an environmental impact committee, refusing to accept a simplified, reductionist account of what is at stake. Particularly with regard to technological advances, the believing community has got to be stubbornly 'from Missouri,' because our account of what is at stake is so much larger. The community must be satisfied with a stressful relationship with the promoters of technological innovations, because we refuse to admit that our truest welfare is served by resolving some sort of malfunction. It might be like clearing up a plumbing problem with dynamite. But, once again, our account always has to be through witness."

Part II: The May Conference

The "New Class" as Guarantor of "the Good Life"

Brigitte Berger

It is the proposition of this paper that the notion of "guaranteeing the good life" is a particular preoccupation of a rising new class of people whose expertise and mode of cognition distinguish them from other groups. As a "new class" of producers and distributors of knowledge, it is concerned with establishing its monopoly over the conceptualization of what constitutes "the good life," just as its many expansionary activities seek to turn the age-old search for "the good life" into entitlements. This shift presupposes that a great variety of interventions are both desirable and necessary. Thus the new-class agenda strongly relies upon legislation, state intervention, adjudication, policing, and punishment on behalf of its class-specific vision, and much time and effort are spent on the propagation of its worldview, on education, and on the raising of the consciousness of a public held to be not yet sufficiently responsive to it.

At this point in time there exists little hard evidence on the precise role of the "new class" in American culture in general and in public policy in particular, although a plethora of impressionistic and anecdotal information is readily available. The purpose of this paper is to provide a rough sketch of the rise and role of the new class and to explore in a preliminary way the linkages between, on the one hand, new-class ideas and interests and, on the other hand, the notion of "guaranteeing the good life" that is about to become dominant in American social life. Inasmuch as the new-class conceptualization of human nature is particularly relevant to the theme of the present meetings, I will make an attempt to identify constitutive elements in the new-class mode of thought, which together with a reliance upon technique and administration make for the peculiarly elastic and unencumbered— if not disembodied—quality of this conceptualization.

In view of the central role I assign to social class in this paper, it may be useful to spell out my understanding of social classes and their dynamics in the process of change and modernization. It is a view that combines a Marxist approach to social class with that of the sociology of knowledge. On the *historical level,* this approach recognizes the importance of economic, political, and structural changes in the modernization process, but it also emphasizes the transformatory power of changes in the structures of the consciousness of individuals, in their modes of cognition, and in their ideas, values, and hopes. So, for instance, while it can be plausibly argued in the Marxist vein that the onset of industrial production dramatically transformed all aspects of social life, including consciousness, it is hard to maintain that this particular "mode of production" appeared suddenly out of nowhere. To the contrary, all the historical materials available strongly suggest that modernity was the result of long—very long—historical processes in the West that included far-reaching changes in human consciousness. Indeed, if industrialization caused great changes in human consciousness, the advent of industrialization itself was rooted in specific prior changes in consciousness such as "individualization" and "rationalization," both of which, if Max Weber was right, may go back as far as the origins of the Judeo-Christian worldview. On the *conceptual level,* my approach to social class presupposes a linkage between the material basis of a given society and the world of ideas typical of it. At the same time, in trying to locate those elements in the *inner dynamics* of the world of ideas, dynamics which in turn relate back to the mode of production, this approach goes beyond the narrow confines of the customary Marxist analyses.

The emphasis upon the inner dynamics of class-specific knowledge is informed by the work of Peter Berger, Brigitte Berger, and Hansfried Kellner, who, in their book entitled *The Homeless Mind: Modernization and Consciousness* (1973), have tried to identify intrinsic components of modern consciousness. They point in particular to the componential quality of knowledge and its peculiar organization, the separability of means and ends, the depersonalization of knowledge and its conceptualization in an abstract frame of reference, continuing "rationalization" and "individualization," and so on, all of which coalesced into a "package of consciousness" peculiar to modern societies regardless of culture and political organization. For the purposes of the argument here, it is important to keep in mind that the package of modern consciousness is more pronounced in the elite sectors of society than in the non-elite ones, and it may be argued that those individuals who and professions which manifest aspects of consciousness and behavior correlative to the modernizing process come to dominate the elite sectors of modern society. Thus elite classes have to be understood as both carriers and manifestations of the modernizing process. They are defined by their specific location in the social structure and the occupational structure, and they reflect class-specific mind-sets and worldviews. However, it is important to keep in mind that a much larger role is assigned to cognition, ideas, and values related to social class, and the contemporary research on the new-class phenomenon presents an important demonstration for the plausibility of this sociology-of-knowledge approach to social classes. As Hansfried Kellner has argued in a recent paper, the new class demonstrates that "*cognition itself* is an important stratifying principle of society."

THE RISE OF THE NEW CLASS AND ITS CONSEQUENCES FOR THE SEARCH FOR "THE GOOD LIFE"

The search for the good life is as old as humankind. Since the dawn of history men and women, individually and collectively, have toiled, schemed, and made extraordinary efforts to attain a good life for themselves and their children. The task of precisely defining what a good life is, however, traditionally fell upon a distinct category of people—for the most part priests and philosophers, frequently a mixture of both. Why that has been so is not difficult to understand. Philosophical anthropologists and sociologists like Max Scheler, Arnold Gehlen, and Peter Berger have argued that human beings beset by the enigma of their own nature and existence must define themselves by referring to what is other than themselves, other than human. For millennia, ordinary people and priests together sought—though on different levels of sophistication—signals of transcendence in a universe that appeared frozen and threatening to them. By the same token, the definition of "the good life" and guidance for its attainment were informed also by reference to the divine rather than by norms generated by the necessities of collective existence. It is important to understand the relationship between culture and the search for transcendence: although this quest by definition can be expressed only in cultural terms and thus must be coached along lines analogous to those of culture and nature—for instance, by making use of images and symbols from either world, the collective (family symbols of father, mother, etc.) or the natural (seasons of growth, maturation, decline)—the religious quest is larger than culture itself. That is, in this vision the roots of religion are *sui generis:* they cannot be reduced primarily either to the psychological needs of individuals or to those of the collective life. In the monotheistic religions this search gave rise to highly sophisticated theologies, which in turn became instrumental in the actual shaping of the course of history.

The broad forces of modernization that have transformed the world with cataclysmic speed brought about a rupture in the embeddedness of the human being in the religious worldview. They also brought to the fore a new group of people, the *new* men of knowledge: natural scientists at first, to be joined at later points in time by technologists, lawyers, and intellectuals. This is not the place to recapitulate the history of intellectuals and the men of science. Yet it is important to note that, until quite recently, natural scientists themselves continued to work largely within a religiously defined frame of reference, as the examples of Galileo, Leonardo, Giordano Bruno, Copernicus, Kepler, Leibniz, and Newton clearly show. It was, of course, widely recognized that the applications of scientific insights lent themselves to a greater control over the adversities of nature. Yet, for quite some time into the modern era, science was understood as providing avenues for the realization of goals of a "higher order." As the natural sciences "matured," however, they became increasingly self-sufficient. In their turning away from an otherworldly orientation toward a "scientific" one, they also became secularized. Pure science, scientists argued henceforth, had a speculative, "riddle-solving" thrust and had to do its own business. It should not be concerned with the implications of its knowledge for either the larger order of things or its immediate practical use. In other words, the collective responsibility of this new breed of scientists related to the enterprise of science only, while the personal responsibility of the individual scientist frequently continued to be informed by reference to the divine.

On the structural level, the modernization process set in motion by this new category of natural scientists also provided the basis for the establishment and expansion of these new groups as professions. However, the astonishing accomplishments of the modern era could not have been achieved without the intervention of two further factors: the rise of technology and the transformation of what Marxists call "the mode of production." The first evolved from an increasing linkage between pure science and its technical application, which produced a new dialectical relationship between theory and practice culminating in a peculiar interactive process between the two (the details of which cannot concern us here). The second, as Max Weber so brilliantly demonstrated, was primarily the result of the rising class of capitalist entrepreneurs who made use of both science and technology in the newly emerging industries. It must always be kept in mind that there exists a functional connection between capitalists, natural scientists, and engineers, who together have to be regarded as important constitutive groups and carriers of the industrial order. As a large number of historical and political studies have amply shown, these three groups need and feed each other in a variety of ways. What has received little attention so far is the fact that the functional connection between these three groups largely explains many of the current tensions that continuously pop up today between the members of the new class and the representatives of the old industrial order, regardless of whether the latter are now owners, managers, scientists, or engineers.

It is of singular importance to understand that the location of the new class in the social structure—and, with this, its peculiar mode of cognition—is fundamentally different from that of this older class of industrial producers. Historically, the new knowledge class has to be seen as a continuation of the old intelligentsia that came into existence at the beginning of the modern era. Broad transformations in the technological, political, and economic structures of Western society provided this group of people with many novel opportunities, a number of which stand out: the significant material progress in all sectors of life, the pluralization of political power, demographic developments leading to urbanization and a general increase in population, and—above all—important technological advancements such as the invention of the printing press, without which modern intellectual production is unthinkable. Although the term "intelligentsia" itself was coined in a different context—that of late nineteenth-century Russia—its use in the present context is nonetheless appropriate. The intellectuals of the modern era make their living by providing nonmaterial services and, more specifically, applying symbolic knowledge, particularly in the areas of education, interpersonal relations, and communication. Not directly dependent upon the modern order of industrial production, yet made possible by the innovations and the very affluence produced by that order, intellectuals came to be a variegated group of individuals. Social scientists like Alfred Weber, Karl Mannheim, and Joseph Schumpeter perceived this intelligentsia for some time as "unattached" to the existing class structure of bourgeois society and lacking in the class-specific interests of the capitalist classes. Because of their apparent independence—they were, in Mannheim's terms, "free-floating"—intellectuals were alleged to be able, if not expected, to detect and castigate the flaws and ills of capitalist society. This mission brought intellectuals into tension with the emergent industrial capitalist order.

While this new category of intellectuals was a small, often marginal group of individuals until well into the nineteenth century (the "coffeehouse" bohemians of Paris and Vienna), the twentieth century saw a phenomenal increase in their numbers. The factors making for this proliferation are many and to a large degree have to be understood as extensions of the forces of modernization identified earlier. Here let me point to the three most obvious. First, numerous astounding technological advances have made for a progressive transformation of the labor force. This transformation has been conducive to ever-further divisions of labor, mostly into uncharted regions of work and life that favor the educated. Second, the growing complexity and opaqueness

of the superstructures of modern civilization have left large segments of the population without knowledge about how to think
and act, and they thus have been propelled to seek guidance from
those claiming competence in ever more narrowly defined fields
of expertise (e.g., the field of counseling recently added its latest
subfield: bereavement counseling). And third, there are important factors flowing from the inner dynamics of professions and
professionalism, such as the monopolistic and imperialistic tendencies of all professions.

These developments taken together have transformed the
category of intellectuals from a marginal, hybrid phenomenon of
capitalist society into a social form that now may be defined in
classical Marxist terms as constituting "a class in and for itself."
The Marxist theoretical requirements—that classes are defined by
their relationship to the means of production; that classes, to be
effective actors in the political arena, must achieve a kind of collective consciousness; and that classes develop distinctive ideologies reflecting their interests—apply to this new collectivity of
intellectuals in a singular manner. For all data available to us
today clearly indicate that the traditional group of intellectuals is
neither marginal nor socially unattached but has coalesced into a
new class. As the new class it controls many of the institutions
central to so-called post-industrial society. In particular, it domi-
nates the modern institutions of the vast educational empire, the
media in all their luxurious forms, and a seemingly boundless
therapeutic apparatus, as well as increasing segments of both
public and private administration.

The public arena of politics provides the primary though not
the only basis for the expansion of the new class. While claiming
independence from the market economy and often defining
themselves in direct opposition to it, members of this new elite
wrap themselves in the mantle of selfless guardians of "individual rights" and define themselves as beleaguered defenders
of an egalitarian social justice. For both purposes they are forced
to rely upon the state in a historically singular manner. In the
process of "operationalizing" these core propositions of their
raison d'être, new-class professionals expand their activities into
more and more areas of public and private life, thereby dramatically enhancing the interventionist potential of the state. In this
fashion the ideological principles—or, if you wish, the cognitive

presuppositions—of the new class ultimately serve its objective class interests. The reliance upon the state and the mechanisms at its disposal is bolstered by a distinctive new-class ideology in which a pronounced adversarial animus against many aspects of bourgeois culture (such as individual responsibility, self-reliance, the desire for advancement, etc.) figures prominently. Instead we now hear a great deal about the "evils of the system" and its abdication from responsibility. Aside from the well-known "triumph of the therapeutic"—which I will say more about shortly—the state now is held to be responsible for virtually every aspect of individual and public life (ranging from questions about the causes of obesity to the issue of eternal peace between nations). This peculiar configuration of factors has pitted the new contenders for power, income, and prestige against the old middle classes, whom they now confront—more or less belligerently and intransigently—in the fight over the hearts and minds of Americans in the political arena. This is hardly what Marx had in mind when he spoke of the inevitability of class struggle in capitalist society, but struggle it is, and class is what it is finally all about.

One of the effects of new-class ideology in action is the removal of the distinction between public and private life, a distinction that has been one of the outstanding consequences of the modernization process. Whereas the old professions, with the exception of the clergy, tended to relate their knowledge and skills primarily to the public sphere and thus had only an indirect influence on individual private life, the proliferating professions of the new class take great pains to establish bridges between the two spheres, and at every turn they seek to transform private matters into public affairs. In this manner the areas of sexuality, health, and the family have been politicized to a degree heretofore unknown in human history. So, for instance, individual sexual preferences have become a matter of individual rights; certain individual patterns of consumption (pertaining not only to nutritional habits but also to habits like smoking) have been declared a public hazard; and in the area of the family a whole array of programs relating to children (prenatal and postnatal care, psychological support as well as day care), to women (the expansion of women's rights into the areas of affirmative action and comparable worth), and to moral education (sex education, values clarification, etc.) have been problematized and transformed from concerns of private life into social issues that now demand government response of one kind or another. What may at one time have been regarded as mechanisms for providing support, aid, and comfort to people in need have now snowballed into entitlement programs. In sum, what we have here is a class of people who regard it as their mission to "guarantee the good life" not only in terms of helping individuals but increasingly in terms of providing public programs, all designed, established, maintained, and—of course—financed via "public" coffers. It may be argued that, in claiming responsibility for the totality of life, the new class most resembles the clergy. It may also not come as a surprise that significant segments of the clergy today have redefined their essential role in ways that are consonant with new-class ideas.

NEW-CLASS MODES OF CONSCIOUSNESS

Like any other social class, this new class too has developed its own class culture and manifests distinctive and highly pronounced class-specific modes of cognition. While this paper is hardly the place to provide a systematic treatise on the variations of knowledge, it is exceedingly important to remember the argument presented in the preamble to this paper: what distinguishes the professions of the new class from those of the old middle classes is not ownership or lack of ownership of the tools of production, but rather their peculiar vision of what constitutes "the good life" as well as the development of mechanisms that will "guarantee" its realization.

Modernity, it has often been said, is characterized by constant innovation, rationality, and reflectivity, and by a corresponding sense of the unreliability and changeability of all social order. At the same time, ever-progressing divisions of labor and technological specialization have confined the ordinary individual to a narrow field of competence. This constellation of factors frequently results in the individual's having a remarkable lack of knowledge about wide spheres of modern life, including the personal and the social. What knowledge the individual does have is necessarily fragmentary, and often is expressed in formulas and clichés. This has grave consequences for the individual. Modern society and its complex superstructures appear increasingly unfathomable to many, and a general feeling of "anomie," a loss of direction, and an uncertainty about how to make sense of individual and collective life—if not, indeed, how to act in it—begin to assert themselves. In this situation, new-class professionals and the expertise they have to offer appear to fill an important gap. Their professional expertise not only appears to provide needed guidance and techniques for the mastery of this or that problem of individual or social life, but also promises salvation from a state of limbo in which many modern individuals appear to be caught. Such broad and generally amorphous expectations resonate well with the self-understanding of new-class professionals. In his various publications Max Scheler wrote about three forms of knowledge—salvational knowledge, educational knowledge, and knowledge for domination—co-existing in all periods of history, including the present. It may well be argued that new-class professionals combine all three in a singular manner.

As pointed out earlier, these professions, like many other categories of modern work, are the product of the increasing division of labor. Thus most categories of modern work share many similarities: status depends upon measurable and certifiable educational attainments, claims to expertise are based upon the application of rationally defined "scientific" theories, and the organization of professional knowledge is guided by logic. At the same time, however, the new-class mode of cognition is fundamentally different from that of the older professions such as science and engineering. This distinction derives primarily from the inner logic of their subject matter and influences the translation of their expertise into social practice in decisive ways. Hansfried Kellner—but also Leon Kass, for instance—have tried to demonstrate that although the field of medicine is certainly a product of the division of labor, medical knowledge itself is grounded in the competent knowledge of the human body and its function (i.e., the scientist is bound to the autonomous logic of his subject). This is hardly the case for many new-class professions. In professions such as those related to pedagogy, social work, varieties of planning therapies of this or that type, social advocacy, and the like, an autonomous logic of competence has been "invented" without a clearly discernible subject matter.

Their claim to competence, Kellner argues, "is frequently based upon a *desire* for competence." Yet these professionals now have to act as if they possess a discrete body of knowledge of indisputable legitimacy. "To further their implied truth claims," Kellner explains, "a number of putative 'professional' theories have been developed for the purpose of common reference. In reality, however, the function of such 'professional' theories is more in the line of self-fulfilling legitimacies, rather than adding to a discrete body of substantive knowledge." In sum, under the guise of rationally defined and organized professionalism, highly subjective contents are peddled.

Professionals, it may be said, can practice their profession only to the extent to which their professional knowledge has not demonstrably been called into question. This assertion typically applies to the knowledge of the natural scientist or the engineer (either a bridge endures or it does not), and all natural-science knowledge is qualified by the "until further notice" clause. It is much more difficult, however, to test the knowledge of new-class professions in the same way. Because of the more or less "contrived" quality of new-class knowledge, Kellner again has argued, the famous Thomas theorem of the "definition of the situation" (which holds that once a situation has been defined as real, regardless of its truth content, it becomes real in its consequences) applies in a most apt manner.

The theoretical deficiency of new-class knowledge produces a number of interesting consequences. The scientific method, it will be remembered, relies not only on observation and measurement but also on the isolation of discrete natural phenomena from the wider structures in which they are embedded, and requires that these be subjected to experimentation under varying conditions. In the transposition of this methodology to the realm of interpersonal relations that the new class considers its own, social phenomena and practices are now isolated as well—that is, they are taken out of their larger context and treated as quasi-autonomous entities. Modern feminism provides a good number of examples of this proclivity. For, according to the new-class perspective, women have to be seen in isolation from the wider structures of personal and social life, including those of the family. This general isolationistic impulse is accompanied by the dynamics mentioned earlier, whereby an already highly abstract mode of cognition advances, as it were, under its own steam, as if in an imaginary space, and can no longer be related back to reality in its immediate sense, though those who follow it step by step can only find it utterly cogent. In describing this process, the terms "hyper-intellectualization" and "hyper-rationalization" suggest themselves.

There is yet a further consequence that needs to be commented on—namely, the lust of new-class professionals for the *problematization* of any and all issues. Everything a lively mind can possibly think of becomes a problem for social engineers who want to apply their expertise. Because the new class has strong vested interests in the multiplication and magnification of problems that may and indeed do exist, it constantly bombards the general public with urgent messages and calls for action on behalf of causes which, needless to say, can be aided only by new-class intervention. The cases of child abuse and sexual harassment may serve as illustrations here. In the actual establishment of intervention programs, however, the traditions of pragmatism and compromise characteristic of the American democratic process tend to resurface. For here new-class conceptualizations are confronted by very different understandings of life. Depending upon how deeply these visions are rooted (as, for instance, in the case of abortion), new-class agendas are derailed or redirected. A good example of the redirection of issues on the new-class agenda can be found in recent developments in which the "politics of women" met with the "politics of motherhood."

In the academy, on the other hand, where no such mechanisms are in place, hyper-intellectualization and hyper-rationalization in all their forms and variations flourish and advance to ever-new heights. At the same time, independent of these intellectual acrobatics, new-class intellectuals in the academy in particular go on to pursue their personal political and economic interests with attitudes that do not reveal nearly the same degree of refinement and sophistication but are, on the contrary, extremely down to earth. The familiar mundane appetites of greed, consumption, domination, self-interest, and the like provide a depressingly predictable counterpoint to the high road of lofty ideals and purposes. The co-existence of these paradoxical factors makes for a situation in which academics now find themselves in need of the very ministrations their professions have created for others.

NEW-CLASS CONCEPTUALIZATIONS OF NATURE AND SOCIAL ENVIRONMENT

The idea of autonomy from biology and nature is an old one. For as long as we can remember, legions of scholars have tried to fathom the parameters of this autonomy. Human beings' actual achievement of greater autonomy from both the natural environment in which they live and their own nature is something that most modern scholars are inclined to ascribe to the powerful effects of science and modern technology. Others, like Peter Berger, accept the idea of an autonomous nature not subject to the gods and their interventions as a *presupposition* of modern science and technology, and trace its origins as far back as the theological thinking of the ancient Israelites. However this autonomy may be explained, the idea of an autonomous individual "who can step outside of his community and even turn against it, is an essential feature of modernity," according to Berger.

During the past five centuries there occurred a progressive diffusion of the rationalistic-technologistic paradigm. In that process the paradigm moved out of the narrow fields of science and technology into ever-wider spheres of life. A straight line appears to run from the control of the external forces of nature and their harnessing for the benefits of individuals and society by means of this paradigm, to the social engineering of collectivities and society itself, including the engineering of the body and psyche of the modern individual. Today the scientific worldview triumphs in all areas of human existence and is inscribed indelibly within the psyche of the modern individual.

Although the convictions of an earlier industrial age are no longer as optimistically expressed today, the same mind-sets persist in the willingness to set goals, to lay plans, to organize and reorganize. The characteristic modes of operation borrowed from the natural sciences and technology consist in attacking the foundations, manipulating the central components, and reconsidering the originating assumptions of whatever one is dealing with. Today the scientific worldview applies to the arts and to society as well as to the conceptualization of human nature, the body, and the psyche.

Analogous to the machine, the body is conceived of as consisting of parts, reparable and replaceable at will. Little can equal modern medicine as a technical triumph and a human contribution. Not only has it controlled innumerable diseases, prolonged life, and substantially improved the general health of sizable portions of humanity, but it also has given us the ability today to control our reproductive functions at will, to interfere with our biological clocks and adapt them to our dreams of life, and—last but not least—to choose our gender at the edge of a surgeon's knife. Getting an ever-firmer control over the body and life itself, human beings are now expected to soar like angels, disembodied and unencumbered by the realities of existence. From the objectified cosmos to the objectified utopian dream, the modern rationalistic-scientific worldview has made for a situation in which human beings seriously believe they can arrive at enduring felicity.

Similar expectations, *mutatis mutandis*, apply to the body social as well. Collective life and its institutions are conceived of as being socially constructed and malleable—they can be deconstructed and reconstructed at will. When the institutions surrounding individuals are changing or are being dismantled, leaving individuals to fend for themselves in a vacuum, they can act rationally only by being egocentric. In recent years Christopher Lasch has made valiant attempts to describe this deconstruction process, which leaves the modern individual in a permanent state of limbo. Because of the pluralistic nature of the modern world, however, the situation is even more complex than Lasch has argued. Modern pluralism flows from factors growing out of the co-existence of a variety of institutional structures (such as those of class, ethnicity, race, and religion) but also from factors growing out of the co-existence of a great many relevance structures (such as those of work, leisure, identity, intimacy, etc.). In the course of a single day, ordinary individuals participate—consecutively as well as concurrently—in different worlds of relevance. So, for instance, an individual may have various concerns about his body—at one moment concentrating on his body's natural needs and functions, at another moment thinking strictly in terms of certain of his body's medical symptoms, and at yet another moment focusing on his body in its relationship to the wider world (the issues of hunger, poverty, pollution, and the like). In the "thematization" of particular relevance structures, individuals increasingly find themselves trapped in a web of confusion and tensions.

Precisely at this point enters the new class. Its paramount professional task is conceived of as providing a balance to these tensions. To this end, new-class professionals seek to uncover new avenues that may lead individuals and society in particular out of impasses created by industrial capitalism and its rationalistic conceptualization of the world.

Like most modern individuals, new-class professionals share in the cognitive presuppositions originating from the natural-science, rationalistic-engineering mode of cognition. These presuppositions have made for a conceptualization of the body, the psyche, and society such that each has its "own rules." However, in contradistinction to the older professions, new-class professions *relate these autonomous rules of body, psyche, and society back to a wider vision of the organismic or "natural wholeness" of all of life.* In the case of medicine, the autonomous rules of the body are now seen to have been repressed, if not endangered, by specific social practices that have developed in the evolution of the modern world. Increasingly, traditional "mainline" medicine (or science) is now perceived as presenting grave dangers, and its "school" practitioners are conceived of more or less as enemies. This fundamental shift between traditional professionals and those of the new class is perhaps nowhere more dramatically expressed than in Ivan Illich's book entitled *Medical Nemesis,* which begins with this memorable sentence: "Modern medicine has become a serious threat to health."

The presupposition that society, the body, and the psyche each has its own rules reveals a further underlying assumption or conviction: namely, that the *true* organism has to be regained again, or at least has to be given the possibility to express itself spontaneously on its own terms.

These two core assumptions of the new class pit it against modern society. However, in their attempts to develop new approaches to social life, the professions of the new class are presented with an analytical paradox when they now ask, What precisely can be done to provide the individual and society with possibilities for spontaneous self-expression? Nature qua nature is incapable of realizing itself; it needs mediation on the part of human actors. In trying to map out new approaches to social life, the new professions are forced to fall back upon the very scientific mode constructed in cognitivistic or rationalistic terms that they sought to bypass. In other words, what is intended to be transcended appears again, simply in different terms. Only the latest paradigm reflects the specific class-loaded notions and interests peculiar to the new class. Thus there exist today two scientific paradigms parallel to each other: one that is openly understood in traditional rationalistic terms, and another that claims to be opposed to it but is nonetheless defined by it. Consequently, in its attempts to engineer spontaneity while opposing the traditional scientific mode of the control of nature and inorganic objects, the new class is propelled to find or rationally construct environments, situations, and social and human relations in order to provide room for the self-expression of a putative organism.

In sum, central to new-class professionalism is not so much an engineering of subject matters, including the human body, as an *engineering of relationships between the body and its environment.* That is to say, new-class logic rests not upon a particular subject matter but rather upon the relationship between an *assumed* inner logic of nature and human life. These new professions tend to accept modern science while they fight it. The assumed inner logic now propels them to focus exclusively on the relationship between organisms and their social relationship, which can be constructed, engineered, and reconstructed. Since these relationships are by definition amorphous and hard to fix, the engineering enterprise is an open-ended one.

If these considerations are now applied to the new-class conceptualization of body and nature, we may observe the following. Whereas the traditional scientist looks at the problems presented by the body as constituting a puzzle that has to be solved in the scientific mode, the new-class professional steps outside of this mode and relates the body back to the larger nature or meaning of things. The traditional scientist is able only to analyze specific symptoms and at best can present his knowledge in terms of "if-then" formulas. He can explicate the possible consequences of his knowledge, but he cannot make decisions for individuals and collectivities. The new-class professional is not bound by such restrictions. If the world does not provide ready-made solutions to a phenomenon identified as a problem, then the world has to be reconstructed to fit the solution defined in new-class visions. Needless to say, there is considerable hubris in what Jacques Maritain has called the "anthropocentric optimism of thought" of this class, which is convinced that changing some of the assumed premises of social existence can bring the world's sufferings to an end. In assuming a salvational mission, the new-class professions may be defined as the priestly engineers of a secularized society.

FROM NEW-CLASS VALUES TO ENTITLEMENTS

The new class has developed its own distinctive value system corresponding to the new cognitive mode just described. At its core stands the value of individual self-realization. This notion is based on the fundamental belief that neither the individual nor social life is a fixed entity but is "makeable" and can be shaped in an infinite variety of ways. When the value of self-realization is connected to the Enlightenment ethos calling for rationality and its extensions (planning, control, and egalitarian justice)—a connection that comes naturally to the new class—we witness the birth of an impulse that *nolens volens* propels this new elite toward political mechanisms that are best described as "entitlements for self-realization."

This peculiar constellation of values and ethos and the consequent dynamics it sets into motion is particularly noticeable in the whole complex of contemporary social policies surrounding the family. Since others more competent than I have commented on this complex of questions, let me here simply adumbrate those moments in the process that serve to illustrate what is at issue here. First, the notion that every individual should realize his or her full potential has placed the modern family under enormous strain and can be held largely responsible for the increased divorce rate. A person decides to marry and to remain committed to a marriage only as long as this value of self-realization is obtainable. Whether or not a woman chooses to abort a fetus depends on her calculations of how much a child will interfere with her goals of self-realization; the care of the child has to be secured in ways that will not prevent the woman from reaching these goals.

The law has accommodated itself to this value. In the legal

context the value of self-realization has been recast in terms of "individual rights." Although jurists in their deliberations tend to weigh this principle against others (such as communitarian principles and the fairness principle), any examination of recent developments demonstrates that the principle of individual rights wins hands down. Indeed, some more philosophically inclined jurists (Ronald Dworkin) have attributed a nearly unimpeachable moral claim to it. In the course of the development of the self-realization/individual rights principle, the law has made certain things progressively easier, among them divorce (a kind of "right to divorce on demand") and abortion (the "right to abortion on demand"). And at this particular point in time we observe intensified agitation on both sides of the political divide for putting child-care programs into place in the form of entitlements as well (a kind of "right to child-care," one may argue).

In recent decades, harsh empirical realities have called into question the feasibility and beneficence of many of the programs and policies created by new-class intellectuals, professionals, and politicians. In fact, a good case can be made for arguing that many of the state programs that intervene on behalf of a great variety of individual and social problems have actually led to the entrenchment—if not the magnification and expansion—of these problems (*vide* the "War on Poverty"). Among those given to honesty and self-reflection there has spread a new soberness about the limits of "doing good." Yet it would be erroneous to expect that the new skepticism would give pause to the imperialistic impulse of the new-class vision of life. Healthy class interests prevent any such developments, and it would take extraordinary resolve and wisdom to reverse a trend that gives credence once more to the Parkinsonian principle of bureaucratization, which holds that "what goes up seldom comes down." What we observe instead is a perpetual widening of the domain of new-class activities, which—in addition to the customary increases in social services, health care, and child-care as well as the stricter application of affirmative-action principles and the like—now includes issues of income security, work laws, full employment, the environment, and, most recently, the balance of trade and payment.

In the course of the expansion of new-class visions of life

that lead to "guaranteeing the good life," another preoccupation of this new elite has been crystallized and has been given urgent status—namely, the preoccupation with the reduction of risks and the guarantee of safety. This concern manifests itself in the demand for risk-free sexuality and the quest for the "flawless child" (a quest which, incidentally, makes the new class particularly dependent upon the new technologies and so-called expert knowledge), just as it manifests itself in global concerns and an intensified pacifism. The search for the risk-free life and a risk-free world reveals a central paradox—namely, that people in our kind of society live better and safer lives than ever before in human history, and yet they are more anxious and paranoid than ever before. Aaron Wildavsky has argued that the biggest risk is to take no risk at all. In the name of safety, new-class agitation has led to policies that control and ultimately curtail any responsible economic growth. After reviewing a large body of empirical data, Wildavsky came to the conclusion that "killing people with safety would be one of the supreme ironies of our time." It does not take great expertise to decide that it is impossible to avoid risk and to see that the notion of a "safe world" is a fundamental illusion. This illusion leads societies into a vicious cycle of activities at the end of which stands the totalitarian state, as Friedrich von Hayek has so convincingly argued. We may be short of the final station on this road, but what has been produced in this manner is a society that is increasingly neurotic, politically impotent, and economically noncompetitive.

In conclusion, we may observe that in the process of transposing a class-specific vision of life to social reality, the new class has made dominant certain mechanisms that lead society a long way down the road to totalitarianism. The new-class vision of life holds sway over much of the surface of society today. In virtually all aspects of their existence, ordinary people—men, women, and children—have become the objects of administrative activities conceived in the new-class mode, and a rapidly acquired technical competence and intellectual flimflam are sufficient to give rational expression to this imbalance in power. In this situation it is a much more difficult, much more personal achievement to express coherently a faith in values that deviate from those currently monopolizing American culture, to confront responsibly the tasks of everyday life that are informed by personal commitments and principles of self-reliance, and to bring oneself to a more nuanced performance of the complex tasks of modern existence.

God, Medicine, and Problems of Evil

Stanley Hauerwas

SUFFERING AND THE GENERATION OF THEODICIES

Dear Ann Landers:

As a health-care professional, I was beginning to think people were becoming more sensible about AIDS, but after reading that letter from "God-fearing and Celibate," the woman who said that sex except for procreation was perverted and AIDS was a punishment from God, I began to feel angry and frustrated again.

I would like to know: Which of God's laws did an 8-year-old hemophiliac break when he contracted the disease from a blood transfusion? And, please tell me, what about the infants born of AIDS mothers? Surely they are not capable of breaking God's laws, yet they, too, have been stricken with this dreadful disease.

Can any sane person believe that God meant for infants to suffer such agony, as infection ravages their tiny bodies and tubes are put in their noses to feed them so they don't die of starvation? Does the writer believe that God intended for infants to wake up in the night screaming from pain caused by the four tubes placed in their chests to drain infected lungs? Where on a baby's chest can you insert four tubes?

213

As a staff nurse in a large pediatric hospital, I have seen several of these infants. I have rocked them in my arms as they cried at night. I have prayed for their recovery, knowing that they will die. In my entire nursing career I have yet to see anything as devastating and unfair as this hideous and unnecessary disease. But never in my darkest moments have I been able to imagine that God, anyone's God, would punish a child in this way simply because his or her mother received a tainted blood transfusion, or used a dirty needle while injecting drugs, or had an affair with a bisexual.

Isn't there enough suffering when one has a terminal illness without adding the guilt of God's disapproval? Please, Ann, tell them one more time.

—Angered in Boston

Dear Boston:

You told them in a way I never could, and I thank you. And now, some good news: For the first time since I have dealt with this subject in the column, I received more letters in support of your point of view than against it. Moreover, as this disease spreads, which is inevitable, we will see a great deal more compassion as it hits our friends, family members and colleagues. You can count on it.[1]

We may well reiterate the question of "Angered in Boston": "Can any sane person believe that God meant for infants to suffer such agony?" We may ask, with Richard Rice, "Why does God let us suffer?"[2] Or we may echo the title of Harold Kushner's book: Why do bad things happen to good people? Given the example of children suffering from AIDS, these questions seem unavoidable. Moreover, it is assumed that people who believe in God are supposed to have or be able to develop answers. Yet I must admit I do not have the heart for such a project. To try to answer such questions—to find ways to justify the goodness of God in the face of an infant dying of AIDS—is not helpful and too often only invites us to play at believing in God. When we are confronted by a dying child, silence seems more appropriate than speech.

1. *Durham Morning Herald*, 12 Apr. 1987, p. 6E.
2. Rice, "The Mystery of Suffering," *Update*, 4 Oct. 1986, p. 3.

However, some cannot remain quiet. For example, Richard Rice suggests a "Christian approach to the problem of suffering," which he says "must pursue three objectives":

> (1) It must affirm the perfect goodness and perfect power of God. A God who is less than perfect in goodness is not worthy of worship, and a God less than perfect in power leaves us without hope. (2) Our response must acknowledge the reality of evil. It is counterintuitive to deny that evil exists, and the view that all suffering is either needed or deserved, removes the negative character of evil. Evil represents an intrusion into God's creation. It is the ultimate absurdity. (3) An adequate response must also provide a basis for meeting it courageously on the level of practical experience.[3]

Rice thinks the "free will defense" best meets these objectives. Evil exists because of a misuse of creaturely freedom. God willed that we be creatures who would serve him out of choice, but as such we can also use our freedom to reject God's authority. God could not create us as free beings who would always use our freedom to do good, for if he did so, then our actions would be determined. "If we ask why God created a world in which suffering was even possible," Rice explains, "the answer is because the highest values of which we know, such as love, loyalty, and compassion, presuppose personal freedom. God cannot create a world where personal values are possible without giving its inhabitants the freedom such values presuppose. All this means there was a risk in creating beings morally free. There was the genuine possibility that they would fall, and this is where evil began. God's creatures, then, are responsible for evil and its consequences, while God is blameless. Because it began in an act of personal freedom, there is no explanation for evil. Indeed evil makes no sense at all."[4]

This is, of course, one of the classic responses to the "problem of evil," but I do not find it logically or existentially compelling. What is such a response meant to do? Is it supposed to console us in the face of a child dying of AIDS? I cannot help but think that such answers, for all their goodwill as well as urgency, are attempts to respond to a question that should never be

3. Rice, "The Mystery of Suffering," *Update,* p. 3.
4. Rice, "The Mystery of Suffering," *Update,* p. 3.

asked. For the answer presupposed is that suffering is some universal experience that raises the same issue at all times. But in fact it is a philosophical and theological mistake to assume that there is a single "problem of evil."

WHY THERE IS NO UNIVERSAL EXPERIENCE OF SUFFERING

I am aware that the claim that there is no universal experience of suffering cannot help but appear odd. Surely, many might argue, *anyone* faced by the agony of a child in pain will see that as suffering. But in fact that is not the case. To see a child's pain as suffering requires that certain habits and practices be in place about the status and place of children. Moreover, often those habits and practices are intelligible only against the background of certain very distinctive beliefs about God and God's relation to us. First let me deal with the issue of the suffering of children.

There is every reason to think that construing a child's pain as innocent suffering is a universal experience among all people. People seem to respond instinctively to a parent grieving over the death of a child. Furthermore, we assume that such suffering binds people in common humanity, that anyone confronted with such misery cannot help but seek explanations and/or causes. Yet descriptively that simply isn't the case. Civilizations, cultures, and communities have existed in which the suffering and death of a child have presented no overriding problem. For example, Colin Turnbull, in his book entitled *The Mountain People*, describes how the Ik would "watch a child with eager anticipation as it crawled toward the fire, then burst into gay and happy laughter as it plunged a skinny hand into the coals. Such times were the few times when parental affection showed itself; a mother would glow with pleasure to hear such joy occasioned by her offspring, and pull it tenderly out of the fire."[5]

Turnbull later reports on the pleasure he felt when he was awakened one night by the mournful wailing one associates with death. He thought this was actually an Ik mother crying over the death of her child. The crying, it turned out, did involve the death

5. Turnbull, *The Mountain People* (New York: Torchbooks, 1972), p. 112.

of a son, but not in the way Turnbull thought. The child's father and mother had argued about whether the child should be buried immediately or the next morning. If they waited, the burial would involve a funeral and require the parents to feast the mourners. The mother refused to bury the child immediately, so her husband beat her and made her dig the burial hole—thus her screams.[6]

Lest it be thought that I am unfairly using this obscure tribe to make my case, let me call your attention to an article entitled "Many AIDS Children Orphaned, Abandoned at Birth," printed a few years ago in the *Durham Morning Herald*. The article notes that about one-third of the children born with AIDS are abandoned at birth.[7] To be sure, some parents abandon their AIDS-afflicted children because they cannot stand the thought of watching them die, but many refuse to care for their children because they fear the disease. Even among enlightened and humane people the fear of death can qualify what we normally assume are our overriding commitments to our children.

The reason I have focused attention on the suffering and death of children is that I assume that in our culture such suffering is the paradigmatic case for the testing of any theodicies. This is particularly the case when we consider the suffering caused by illness. For despite our sophistication about the nature and causes of illness, many of us continue to interpret illness as a punishment for or at least the result of our living as we should not. What so often bothers us about illness is not simply the pain and suffering it occasions but also its sheer irrationality. If we are able to make the disease a consequence of a way of life—that is, attribute it to smoking or a stressful lifestyle—it is almost comforting. We may be sick, but at least we know *why* we are sick. This, of course, is but a form of the free-will defense: we can attribute the evil we suffer to our agency—we got what we deserved.

This is what we cannot do in the case of childhood illness. No doubt some children are sick because of the actions of adults—legal and illegal drug use during pregnancy, air pollution—but the fact that children suffer and even die from such causes makes their deaths no less tragic. Yet most childhood diseases can't be

---

6. Turnbull, *The Mountain People*, pp. 129-30.
7. "Many AIDS Children Orphaned, Abandoned at Birth," *Durham Morning Herald*, 4 Sept. 1987, p. 18A.

blamed on anyone. The sick child simply got the bad luck of the draw. There is no "causal" explanation that can remove the absurdity of a child's death and suffering. Most of us surely believe that any account which describes a child's suffering in terms of desert cannot be anything less than obscene.

DISTINGUISHING THE PRACTICAL ISSUES FROM THE THEORETICAL QUESTION

In our culture the suffering of a child seems to challenge our most fundamental beliefs about the goodness of God as well as the meaningfulness of human existence because such suffering seems pointless. It generates questions like the one posed by Richard Rice—"Why does a God perfect in power and goodness let such suffering happen?" But to ask the question in this way is to transform what is essentially a practical issue requiring a practical response into a theoretical question of only speculative interest.

The theoretical question is of relatively recent origin. To ask the question theoretically—that is, How can a good God allow evil?—depends on the assumption that we can know what it means for God to be good apart from any community's ongoing life. God and goodness are theoretical constructs that can be shown to exist by necessity on the grounds of anyone's reason. What it means for God to be powerful and/or good requires no context of the cross and resurrection of Jesus of Nazareth, but rather can be known in and of itself. As I shall try to show, this way of asking the question is generated primarily by the epistemological assumptions of the Enlightenment and, if accepted by Christians, can only pervert our best insights concerning the nature of suffering as well as our response to it.

In particular, when the problem of evil is presented as a theoretical question that can be asked by anyone, the most important element is lost—namely, the identity and historical situation of the one who asks the question. Douglas John Hall rightly reminds us that "no human question is ever asked (and no answer given) in a historical vacuum; it is asked in a specific time and place by specific persons. With certain kinds of questions this contextual dimension may not be so significant, but with our present question it is of primary importance. The aspects of the problem of suffering which we shall hold up, as well as the responses that we shall give to them, will be determined in great measure by the particular circumstances, openly acknowledged or silently assumed, in which we find ourselves."[8]

What we must recognize is that one of the reasons that modern discussions of the so-called theodicy question fail to help us in any way that matters is that they are abstracted from any concrete historical context. No one has argued this more powerfully than Ken Surin in his *Theology and the Problem of Evil*. Following Alasdair MacIntyre, Surin notes that the seeming contradiction between benevolent divine omnipotence and the existence of evil did not make Christian thinkers of the past consider the existence of God unintelligible. MacIntyre suggests that the apparent incoherence of evil and belief in God was tolerable because religious faith was a set of skills that was indispensable for making intelligible descriptions used in social and intellectual life.[9] In other words, suffering and evil were aspects of existence anticipated by a community schooled by a narrative that provided a means to go on even if evil could not be "explained." Indeed, it was crucial that suffering not be "explained" because such explanation would undercut the necessity of the community designed to respond to such suffering. That pointless suffering seems to render God's existence problematic for us indicates that something decisive has happened in our history, has happened to our understanding of God as well as our understanding of ourselves.

In an effort to help us understand what has happened, Surin calls attention to Augustine's treatment of evil. According to Surin, Augustine attributes the intractable quality of evil to a human propensity that is driven ever deeper by habit and whose destructive power is enhanced by memory. "The memory's capacity to reinforce illicit pleasures is accompanied by a seemingly inexhaustible perversity which impels the individual to repeat, and thus to remember ever more vividly and insistently, the delight gained from every wicked deed of the past" (*TPE*, p. 10). According to Augustine, therefore, we are hopelessly con-

8. Hall, *God and Human Suffering* (Minneapolis: Augsburg, 1986), p. 24.

9. Surin, *Theology and the Problem of Evil* (Oxford: Basil Blackwell, 1986), p. 9. All subsequent references to this volume will be made in the text as follows: *TPE*, p.___.

strained by a memory habituated by wickedness so that the "problem of evil" cannot be solved by philosophical maneuvers.

Rather than use this approach, we must see the problem against the background of the goal of every Christian: the attainment of blessedness, the way God has made possible through Jesus Christ. Ironically, the problem of evil can be located rightly when we understand that we are beings who were created to enjoy our status as creatures but who refuse to accept the gifts necessary for such joy. Thus the only hope for a resolution to our evil lies in the free gift of a gracious God. As Surin says, "So it is conversion—which comes about when the human will cooperates with divine grace—that solves the 'problem of evil.' The unconverted person's endeavors to resolve 'the problem of evil,' no matter how sincere and intellectually gifted the person might be, are doomed ultimately to be self-defeating. Only faith in Christ makes possible the cleansing of our vision, a cleansing regarded by Augustine as the necessary preliminary to the vision of God" (TPE, p. 11). Without such conversion, the very process of seeking to answer the question of evil will only invite self-deception and further sin.

According to Surin, Augustine's account of evil and the process of conversion must be supplemented by taking account of Augustine's historical context. Augustine, after all, was writing soon after Constantine legalized the Christian religion, a time when Christians were no longer being martyred but were enjoying positions of privilege. Accordingly, enemies of Christianity would no longer be "without"; they would be "within." Moreover, the "problem of evil" so conceived must be seen against Augustine's theology of history, which attempts to see the whole of history as the outworking of God's providence (TPE, p. 12).

Whatever flaws Augustine's account of evil has, and I think it has many, Augustine nonetheless sees the problem of evil as a practical problem. In this respect he continues to think of suffering from a Pauline perspective—as an opportunity for living in a way more faithful to the new age. Thus Paul can say in Romans 5, "Therefore, since we are justified by faith, we have peace with God through our Lord Jesus Christ. Through him we have obtained access to this grace in which we stand, and we rejoice in our hope of sharing the glory of God. More than that, we rejoice in our sufferings, knowing that suffering produces endurance, and endurance produces character, and character produces hope, and hope does not disappoint us, because God's love has been poured into our hearts through the Holy Spirit which has been given to us" (vv. 1-5). Apparently it never occurred to the early Christians to question their belief in God or even God's goodness because they were unjustly suffering for their beliefs. Rather, their faith gave them direction in the face of persecution and general misfortune. Evil was not a metaphysical problem needing solution but a pressing practical challenge requiring an immediate response.

Therefore, the so-called problem of evil is not and cannot be a single problem, for it makes all the difference what God one worships as well as how one thinks God is known. In contrast to Augustine's understanding of the problem of evil, the modern understanding of it is that it entails a theoretical enterprise which entertains the possibility of the existence of an omnipotent God. As Surin suggests, the subject that now asks the question about evil "is the putatively rational and autonomous individual who confines herself [or himself] to the entirely *worldly* discipline of 'evidencing' and 'justifying' cognitive formations which, moreover, are restrictively derived from reason and sense-experience. This worldly discipline, which finds its authoritative manifestation in common-sense rationalism and empiricism, would cease to be what it essentially is if it were required to posit a subject whose self-definition required her [or him] to live and think as 'servant of God'" (*TPE*, p. 13).

Theodicy done in this theoretical mode abstracts the so-called problem of evil from any theological account of history. Instead, Surin points out, theodicy is seen "as an ahistorical and individualistic quest for logically stable notions, exact axioms, and rigorous chains of deductive inference. Unlike Augustine, the post-Leibnizian theodicist does not feel constrained to understand history as anything possessing an intrinsic *thematic* importance, let alone as a history which is the work of the very God who reveals himself in Jesus Christ, and which is a determining element in the subject's self-definition" (*TPE*, p. 13). The "philosophical theism" which generates the assumption that there is a thing called the "problem of evil" has little in common with the attitude of pre-Enlightenment theologians, who saw the problem of suffering as a practical challenge for the Christian community.

For Christians, suffering—even the suffering of a child—cannot be separated from their calling to be a new people made holy by conversion. For conversion is "inseparable from fellowship, a fellowship which is at root fellowship with . . . the Trinity itself. This fellowship is inseparable from commitment to a community, a commitment expressed in sharing its way of life, its customs and practices" (*TPE*, p. 23). So Christians have not had a "solution" to the problem of evil. Rather, they have had a community of care that has made it possible to absorb the destructive terror of evil which constantly threatens to destroy all human relations.

It is clear that something has gone decisively wrong for Christians when we underwrite the widespread assumption that there is a so-called problem of evil which is intelligible from anyone's perspective—that is, when we turn the Christian faith into a system of beliefs that can be or is universally known without conversion to a life within a specific community of people. In effect, it is to underwrite the Enlightenment assumptions that we are most fully ourselves when we are free of all tradition and community other than those we have chosen from the position of complete autonomy. In such a context, suffering cannot help but appear absurd, since it always stands as a threat to that autonomy.

Lest it be thought that I am being particularly hard on the epistemological assumptions of the Enlightenment, let me say that I think what the great Enlightenment thinkers often did was to provide a secular account of Christian tendencies. To make this case, I will need to make some unguarded generalizations as well as historical judgments that would require greater nuance to defend. However, since I think it better to be interesting than to be right, I'm going to storm ahead. Moreover, by proceeding in this way I hope to make clear the peculiar dilemma in which medicine finds itself.

Surin rightly criticizes modern theodicies which assume foundational epistemologies that attempt to solve the problem of evil from some ahistorical standpoint. Yet this tendency, I suspect, is but the continuing habit of mind schooled by Christian pretensions that our beliefs are explanatory accounts sufficient to show why "the way things are" is such by necessity. I think this habit of mind developed when Christianity became a civilizational religion oriented to provide the ethos necessary to sustain an empire. Rather than being a set of convictions about God's work in Jesus Christ requiring conversion and membership in a community, Christianity became that set of beliefs which explains why the way things are is the way things were meant to be for any right-thinking person, converted or not.

There is a definite sociological correlative to this understanding of Christianity. It is the seeming correlative between the task of the church and the task of the state. Christians, rather than trying to survive and endure in often hostile political contexts, now have a stake in insuring that the policies of the Christian emperor will be successful. In order to do so, we must see history as a sphere of strict cause and effect, and thus exclude chance as much as possible from any explanatory account of history. In such a context, evil appears as that which has not yet come under our control. Crucial to this set of assumptions is the assumption that the good must ultimately triumph; otherwise the universe as well as the social order is incoherent. In such a context, the "problem of evil" is—to put it crudely but I think accurately—the challenge to show why those with the right beliefs do not always win in worldly terms. Theodicy in the theoretical mode, which is acutely criticized by Surin, is but the metaphysical expression of this deep-seated presumption that our belief in God is irrational if it does not put us on the winning side of history.

This explains the extraordinary presumption that a crisis of faith is created by "bad things happening to good people." Here a mechanistic metaphysic is combined with a deistic account of God; in this way the pagan assumption that god or the gods are to be judged by how well it or they insure the successful outcome of human purposes is underwritten in the name of Christianity. It is assumed that the attributes of such a god or gods can be known and characterized abstractly. The God of Abraham, Isaac, and Jacob is not the god that creates something called the "problem of evil"; rather, that problem is created by a god about which the most important facts seem to be that it exists and is morally perfect as well as all-powerful.

WHY "ANTHROPODICY" HAS REPLACED THEODICY

It is certainly true that the theoretical god of theoretical theodicies has made many concerned about the problem of suffering. Yet I think it is also the case that the attempt to defend such a god in the face of human suffering is largely irrelevant for the lives of most people actually doing the suffering. This is so because the god that such theodicies try to justify is, by definition, largely irrelevant to most people's lives. I am well aware that pollsters tell us that over 90 percent of the American people believe in god, but it is extremely unclear what such a claim means. I suspect the god in which most of us believe, or at least the god with which we live our lives, is the god we find in theoretical theodicies that is unable to call into existence a people who can provide an alternative to the world. The true god has been driven from the world by the assumption that we must control our existence by acquiring the power to eradicate from our lives anything that threatens our autonomy as individuals.

Ernest Becker has been one of the most insightful commentators on the implications of this view of the world, which he characterizes as a product of the Newtonian revolution. According to Becker, it was Newton's great achievement to rehabilitate nature as having human significance—that is, to show that a nature which is not animated with godly purpose can nonetheless serve human ends. But this rehabilitation came at a cost, as Becker explains:

> If the new nature was so regular and beautiful, then why was there evil in the world? Man needed a new theodicy, but this time he could not put the burden on God. Something entirely different had to be done to explain evil in the world, a theodicy without Divine intervention. The new theodicy had to be a natural one, a "secular" one. The challenge was all the greater because the human mind was not prepared for such ingenuity: the idea of a "secular" theodicy was a contradiction in words and in emotions. Yet it describes exactly what was needed: an "anthropodicy." Evil had to be explained as existing in the world apart from God's intention or justification. Furthermore, as God was gradually eliminated from science as an explanatory principle, the need for a complete theodicy also finally vanished. There could be no sensible explanation for *all* the evil to which life is subject, apart from a belief in God—certainly no explanation that mere mortals could attain. Consequently, man had to settle for a new *limited* explanation, an anthropodicy which would cover only *those evils that*

allow for human remedy. The only way to achieve this new explanation was gradually to shift the burden from reliance on God's will to the belief in man's understanding and powers. This was a shift that was to occupy the whole Enlightenment, and it was not accomplished easily. In fact, it is still not accomplished today.[10]

If Becker is right that the real problem is how to account for those evils which allow for human intervention, we can better understand why illness and, correlatively, medicine seem to become the context for raising the question of evil. As moderns we are bothered by the evil effects of hurricanes, tornadoes, and volcanoes, but they do not trouble us the way sickness does. If we are injured in a natural disaster, we think that's bad luck. Such evil really requires no general explanation because there is nothing we can do or could have done to prevent it. Once we no longer believe in the God of creation, there is no god that such disasters call into question.

But sickness is quite another matter. Sickness should not exist because we think of it as something in which we can intervene and as something that we can ultimately eliminate. Sickness challenges our most cherished presumption that we are or at least can be in control of our existence. Sickness creates the problem of "anthropodicy" because it challenges our most precious and profound belief that humanity has in fact become god. Against the backdrop of such a belief we conclude that sickness should not exist.

In such a context medicine becomes the mirror image of theoretical theodicies sponsored by the Enlightenment as it attempts to save our profoundest hopes that sickness should and can be eliminated. We must assume a strict causal order so that this new emperor can be assured of success. We do not need a community capable of caring for the ill; all we need is an instrumental rationality made powerful by technological sophistication.

The ideology that is institutionalized in medicine requires that we interpret all illness as pointless. By "pointless" I mean that it can play no role in helping us live our lives well. Illness is an ab-

10. Becker, *The Structure of Evil* (New York: George Braziller, 1968), p. 18.

surdity in a history formed by the commitment to overcome all evils that potentially we can control. I suspect that this is one of the reasons we have so much difficulty dealing with chronic illness—it should not exist but it does. It would almost be better to eliminate the subjects of such illness rather than have them remind us that our project to eliminate illness has made little progress.

It is only against this background that we can appreciate the widespread assumption that we ought to do what we can do through the office of science and medicine. Whereas it used to be a physician's first obligation *not* to act, we now believe our commitment to the abolition of limits makes the physician's first obligation *to act* through the office of medicine. As a result, physicians lose their freedom to care for the sick because they are now judged by the predictability of their performance. The patient becomes the consumer, and thus the old conception of medicine as a collaborative enterprise, in which doctor and patient each have freedoms and responsibilities, can no longer be sustained."[11]

One of the most disturbing aspects of this change is the assumption that a medicine so formed needs no justification. We assume we ought to do what we can do because such is the way of compassion. But as Oliver O'Donovan reminds us, compassion is hardly a virtue that can stand by itself. For compassion is a motivating virtue that responds to suffering, to the infringement of passive freedom, by prompting us to action, thus circumventing thought. It therefore presupposes that an answer has already been given to the question "What needs to be done?"[12] When compassion becomes the overriding virtue in a world in which we believe that human inventiveness has no limits, the results can only be the increasing subjection of our lives to a technology grown cruel by its Promethean pretensions. Thus, ironically, the evil we now suffer is the result of our fevered attempts to avoid the reality that we will always live in a world where some children will die. In an attempt to "solve the problem of evil," we cannot help but suffer from the results of failing to recognize that it cannot be solved if by "solved" we mean "eliminated."

11. Oliver O'Donovan, *Begotten or Made?* (Oxford: Oxford University Press, 1984), pp. 1-10.

12. O'Donovan, *Begotten or Made?* p. 11.

WHY WE STILL CANNOT AVOID THE SUFFERING OF CHILDREN

Even if I have successfully convinced you that it's a mistake to try to "solve" the problem of evil, particularly in our circumstances, we still confront children who are suffering and possibly dying. I have no answer to the question of why children suffer because I don't know what the question means. I do, however, have a suggestion about why we are bothered by the suffering of children. When illness breaks into our lives, it subverts our plans and projects. As adults we may respond more or less well to our illness, but at least the illness seems to have a context—we can make it part of our ongoing story. Indeed, it may well be that one of the most valuable functions of medicine is to help us go on in the face of an illness that may not finally be curable.

In the preface of his whimsically titled book, *The Man Who Mistook His Wife for a Hat*, Oliver Sacks stresses the importance of making a case history a personal story:

> Animals get diseases, but only man falls radically into sickness.
>
> My work, my life, is all with the sick—but the sick and their sickness drive me to thoughts which, perhaps, I might otherwise not have. So much so that I am compelled to ask, with Nietzsche: "As for sickness: are we not almost tempted to ask whether we could get along without it?"—and to see the questions it raises as fundamental in nature. Constantly my patients drive me to question, and constantly my questions drive me to patients—thus in the stories or studies which follow there is a continual movement from one to the other.
>
> Studies, yes; why stories, or cases? Hippocrates introduced the historical conception of disease, the idea that diseases have a course, from their first intimations to their climax or crisis, and thence to their happy or fatal resolution. Hippocrates thus introduced the case history, a description, or depiction, of the natural history of disease—precisely expressed by the old word "pathography." Such histories are a form of natural history—but they tell us nothing about the individual and *his* history; they convey nothing of the person, and the experience of the person, as he

faces, and struggles to survive, his disease. There is no "subject" in a narrow case history; modern case histories allude to the subject in a cursory phrase ("a trisomic albino female of 21"), which could as well apply to a rat as a human being. To restore the human subject at the centre—the suffering, afflicted, fighting, human subject—we must deepen a case history to a narrative or tale: only then do we have a "who" as well as a "what," a real person, a patient, in relation to disease—in relation to the physical.[13]

Ironically, with the instrumentalization of modern medicine, the patient is seen as a passive recipient of medical expertise, and he or she can no longer be seen as a person with a life story. "Yet it is only within a life story," notes David Barnard, "that illness has a meaningful place. And to see the patient's illness as a development in a biography—rather than as an isolated series of biological events—is precisely to recover a context of meaning for medical interventions. Within that context, it is possible to appreciate both the patient's own values and hopes, which supply a framework for evaluating specific medical interventions, and his or her reservoirs of agency and strength, which are necessary complements to physician power in healing relationships."[14]

I think childhood suffering bothers us so deeply because we assume that children lack a life story which potentially gives their illness some meaning. In that respect I suspect we often fail to appreciate the richness of their young world as well as their toughness and resilience. But I suspect what bothers us even more about childhood suffering is that it makes us face our deepest suspicions that all of us lack a life story which makes us capable of responding to illness in a manner that enables us to go on as individuals, as friends, as parents, and as a community. I suspect that if Christian convictions have any guidance to give us about how we are to understand as well as respond to suffering, it is by helping us find that our lives are located in God's narrative—the God who has not abandoned us even when we and those we most care about are ill. But to develop that suggestion is another story.

13. Sacks, *The Man Who Mistook His Wife for a Hat and Other Clinical Tales* (New York: Harper & Row, 1985), pp. xiii-xiv.

14. Barnard, "Religion and Medicine: A Meditation on Lines of A. J. Heschel," *Soundings* 68 (Winter 1985): 456-57.

Guaranteeing the "Quality" of Life through Law: The Emerging Right to a "Good" Life

Robert A. Destro

> *A new age will come which from the standpoint of a higher morality will no longer heed the demands of an inflated concept of humanity and an overestimation of the value of life as such.*
>
> Alfred Hoche, 1920[1]

> *We can no longer base our ethics on the idea that human beings are a special form of creation, made in the image of God, singled out from all other animals, and alone possessing an immortal soul. Our better understanding of nature has bridged the gulf that was once thought to lie between ourselves and other species, so why should we believe that the mere fact that a being is a member of the species Homo sapiens endows its life with some unique, almost infinite, value?*
>
> Peter Singer, 1983[2]

1. Hoche, "Ärztliche Bemerkungen," in K. Binding and A. Hoche, *The Permission to Destroy Life Unworthy of Life (Die Freigabe der Vernichtung lebensunwerten Lebens: Ihr und ihre Form)*, 1920, pp. 61-62.
2. Singer, "Sanctity of Life or Quality of Life?" *Pediatrics* 72 (July 1983): 128.

INTRODUCTION

The long-standing academic and professional debate over whether "quality-of-life" or natural-rights ("sanctity-of-life") ethics should govern the resolution of bio-ethical questions has entered a new and critical phase: the legal one. The debate is not new for either bio-ethics[3] or the law.[4] But to the extent that quality-of-life theories remain in the realm of polite academic discourse, public policymakers will be called upon (as they rarely are) to face the significant issues such a dichotomy presents for them. Should the law itself adopt the view that the legal protection of an individual's life is to be determined by its *quality* rather than by its human *nature*, it would be an understatement to assert that the impact would be profound.

This conference thus comes at a propitious time; for even as courts and policymakers are increasingly called upon to decide which ethic shall be their guide, there remains within the public-policy debate a distressing tendency to rely on polite legal fictions and medical jargon to avoid direct discussion of the central ethical question.[5] And what is that question for the law? It is this:

> Should the prohibitions of the criminal and civil law be relaxed to permit active and passive steps to be taken with the specific intent of ending one's own or another's life when that life has lost meaning, quality, or value to oneself or others?

At a more fundamental level, the question is this: At what point during the continuum of biological human existence from con-

---

3. See generally Alexander, "Medical Science under Dictatorship," *New England Journal of Medicine* 241 (1949): 39; R. Lifton, *The Nazi Doctors: Medical Killing and the Psychology of Genocide* (New York: Basic Books, 1986); D. Kevels, "Annals of Eugenics: A Secular Faith," *New Yorker,* 8 Oct. 1984, at 51 (Part I); 15 Oct. 1984, at 52 (Part II); 22 Oct. 1984, at 51 (Part III); 29 Oct. 1984, at 92 (Part IV).

4. See, for example, *Skinner v. Oklahoma,* 316 U.S. 535 (1942) (sterilization of certain criminals); *Buck v. Bell,* 274 U.S. 200 (1927) (sterilization of the mentally incompetent); *United States v. Altsoetter* (Military Tribunal III, 1947) ("The Justice Case"; eugenics courts).

5. See, for example, *In re Guardianship of Grant,* 109 Wash.2d 545, 747 P.2d 445, 449, 451 (1987) (an incompetent patient can "refuse" treatment; withholding nutrition and hydration from those not in the process of dying is not passive euthanasia).

ception until death should the law impose a *duty* to protect and preserve an individual's life?[6]

In this paper I will focus on the degree to which the current direction of American case law has been affected by the view that legal protection should not attach when a life is deemed to have no value or meaning. After an examination of legal method, I will focus briefly on three main points: individual autonomy, surrogate decision-making, and the degree to which a "quality-of-life" ethical approach has influenced the direction of current legal policy.[7] I will conclude with an argument that it is now well past the time for those who are charged with the duty to provide moral and ethical guidance in political and religious forums to provide it.

I. FROM THE "RIGHT TO LIFE" TO THE "RIGHT TO A GOOD LIFE": LEGAL REASONING AND PROCESS IN QUALITY-OF-LIFE CASES

A. Introduction

Overt discussion of quality-of-life issues in reported cases is of relatively recent vintage. Rapid advances in medical technology have accelerated the process, but the quality-of-life *direction* of the debate was clear with respect to abortion[8] and euthana-

---

6. Professor Philip G. Peters, Jr., of the School of Law of the University of Missouri, Columbia, has raised the "duty to protect" issue in an article entitled "Protecting the Unconceived: Nonexistence, Avoidability, and Reproductive Technology," *Arizona Law Review* 31 (1989): 487.

7. For a more extensive discussion, see generally R. Destro, "Quality of Life Ethics and Constitutional Jurisprudence: The Demise of Natural Rights and Equal Protection for the Disabled and Incompetent," *Journal of Contemporary Health Law & Policy* 2 (1986): 71.

8. See, for example, Colo. Rev. Stat., Sec. 40-2-50(4) (1963 & Supp. 1967), 1967 Colo. Sess. Laws, Ch. 190 at Sec. 284 (permitting eugenic abortions); *Gleitman v. Cosgrove*, 49 N.J. 22, 227 A.2d 689 (1967) (rejecting the tort of "wrongful birth" based on the birth of a child with a disability); Model Penal Code, Sec. 207.11 (1959) (permitting eugenic abortions). See generally the National Commission for the Protection of Human Subjects of Biomedical and Behavioral Research, *Report and Recommendations: Research on the Fetus* (1975), reprinted in *Fed. Reg.* 40 (1975): 33, 530 (partially codified at 45 C.F.R., Sec. 46.101-.301 (1976); Symposium, "The Report and Recommendations of the National Commission for the Pro-

sia[9] long before the courts were called upon to grapple with the advances in medical technology so often cited as current justification for a shift in our ethical approach.[10]

Even today, consideration of the quality-of-life issue largely takes place between the lines of cases involving the rights of the disabled and of sick adults and infants to food, water, and basic medical care. The discussion is more overt in cases raising what have become known as "wrongful birth" and "wrongful life" claims, but these are the exception rather than the rule.[11] To be sure, the central ethical dilemmas are recognized, but the premium is placed upon adopting a compassionate, pragmatic or "problem solving" approach appropriate to resolving the case before the court.[12] A disclaimer to the effect that the court majority

---

tection of Human Subjects of Biomedical and Behavioral Research: Research on the Fetus," *Villanova Law Review* 22 (1977): 297-417; R. Morison, "Implications of Prenatal Diagnosis for the Quality of and Right to Human Life," in *Biomedical Ethics and the Law*, ed. J. Humber and R. Almeder (New York: Plenum, 1979).

9. See, for example, *Death, Dying, and Euthanasia*, ed. D. Horan and D. Mall (Frederick, Md.: University Publications of America, 1977); D. Humphry, *Let Me Die Before I Wake* (Los Angeles: Hemlock Society, 1981); *Beneficent Euthanasia*, ed. M. Kohl (Buffalo, N.Y.: Prometheus Books, 1975); P. Ramsey, *Ethics at the Edges of Life* (New Haven: Yale University Press, 1978); Sacred Congregation for the Doctrine of the Faith, *Declaration on Euthanasia* (Vatican City, 5 May 1980); Alexander, "Medical Science under Dictatorship," p. 39; J. Rachels, "Active and Passive Euthanasia," *New England Journal of Medicine* 292 (1975): 78; Y. Kamisar, "Euthanasia Legislation: Some Non-Religious Objections," *Minnesota Law Review* 42 (1958): 969.

10. See generally J. Lyon, *Playing God in the Nursery* (New York: W. W. Norton, 1985); *How Brave a New World*, ed. R. McCormick (New York: Doubleday, 1981); President's Commission for the Study of Ethical Problems in Medicine and Biomedical and Behavioral Research, *Deciding to Forego Life-Sustaining Treatment* (Washington: GPO, 1983); R. Scott, "Legal Implications and Law-Making in Bioethics and Experimental Medicine," *Journal of Contemporary Health Law & Policy* 1 (1985): 47.

11. See the text accompanying notes 79-84 herein. Professor Peters' article called "Protecting the Unconceived" (see n. 6) contains a thorough and thoughtful analysis of the "nonexistence test" (the idea being that nonexistence is preferable to life with certain disabling conditions) as a basis for state intervention in parental reproductive decision-making on behalf of their as-yet-to-be-conceived children.

12. See, for example, *Rasmussen v. Fleming*, 154 Ariz. 207, 741 P.2d

"emphasize[s] that [they] are not endorsing suicide or euthanasia" or that they are explicitly rejecting quality-of-life reasoning[13] is usually noted for the record.[14] Nevertheless, it is clear from the discussion that were the quality of the patient's life not in serious question, death would not be considered to be a legitimate option.[15] It thus falls to the dissenters to discuss the difficult ethical issues on their merits.[16]

674, 686 (1987) (referring to "standards of logic, morality and medicine" as affecting judgments about the treatment of the terminally ill), quoting *Eichner v. Dillon*, 73 A.D.2d 431, 464-65, 426 N.Y.S.2d 517, 542-43 (App. Div. 1980); *Matter of Beth Israel Medical Center*, 519 N.Y.S.2d 511, 514 (Supp. 1987) ("essentially one of the most difficult social and ethical issues of our times").

13. *In re Gardner*, 534 A.2d 947, 955 (Me. 1987).

14. *In re Guardianship of Grant*, 109 Wash.2d 545, 747 P.2d 445, 454 (1987); in accordance with *Brophy v. New England Sinai Hospital, Inc.*, 398 Mass. 417, 497 N.E.2d 626, 635 (1986); *Matter of Beth Israel Medical Center*, 519 N.Y.S.2d 511 (Supp. 1987); *Rasmussen v. Fleming*, 154 Ariz. 207, 741 P.2d 674 (1987); *Delio v. Westchester County Medical Center*, 516 N.Y.S.2d 677 (A.D. 2 Dept., 1987); *Matter of Peter*, 108 N.J. 365, 529 A.2d 410 (1987); *In re Jane Doe*, 533 A.2d 523 (R.I. 1987); *Gilbert v. State*, 487 (Fla. App., 4th Dist., 1986).

15. Take, for example, the majority's approach in *In re Guardianship of Grant*, 109 Wash.2d 545, 747 P.2d 445, 455 (1987):

> The prolongation of the existence of [Barbara Grant's] vegetative state for possibly years to come by artificially placing liquids and nutrients into this body to the emotional and economic destruction of the survivors is a monstrous assault to the family concerned that we will not countenance.

Now compare it with that of two of the dissenting justices:

> The fact that her functioning is limited does not mean that it is in her best interest to die. I object to the majority's obvious judgment that Barbara's life is not of value. I believe it is inappropriate for the majority to authorize a guardian to determine that the ward's life is not worth living because the guardian deems that "life" to be negligible. The increasing trend of courts to decide that life of "lesser" quality is not worth living is quite disturbing. (at 463)

(Goodloe and Dore, JJ., dissenting).

16. See, for example, *Brophy v. New England Sinai Hospital, Inc.*, 398 Mass. 417, 497 N.E.2d 626, 640 (1986) (Nolan, Lynch, and O'Connor, JJ., dissenting in whole or in part, respectively; 4-3 decision); *In re Gardner*, 534 A.2d 947, 956 (Me. 1987) (Clifford, Roberts, and Wathen, JJ., dissenting; 4-3 decision); *In re Guardianship of Grant*, 109 Wash.2d 545, 747 P.2d

B. The Role of Individual Autonomy

The rights of the individual receive significant protection under American law. Constitutional law, for example, provides guarantees against the invasion of certain rights by government, tort law provides compensation for breaches of civil duty that cause harm to persons or their interests, and family law (including certain aspects of the law of decedents' estates) provides a framework that sets forth the reciprocal interests and duties of individuals and their kin.

In the context of the present discussion, principles drawn from each of these fields of law play an important role in shaping the courts' approach to cases involving quality-of-life issues. It will therefore be necessary to explain briefly a few important distinctions.

1. The Constitutional "Right to Privacy"

The first of these distinctions involves constitutional law and is often described as "the right to privacy." Although the term "privacy" does appear in several state constitutions,[17] it does not appear in the U.S. Constitution. As commonly understood in constitutional terms, the term "privacy" refers to two distinct concepts: (1) the inviolability of one's *person, home,* or *things* from unreasonable governmental intrusions; and (2) individual autonomy or liberty with respect to certain matters important to one's person or the course of one's life (e.g., marriage, sex, childbearing). The protection for the locational aspect of privacy is found in the Fourth Amendment,[18] whereas the Due Process Clause of the Fourteenth Amendment[19] is generally held to be the basis of

---

445 (1987) (Andersen, Brachtenbach, Goodloe, and Dore, JJ., dissenting in whole or in part, respectively; 5-4 decision).

17. See, for example, the constitution of Arizona, art. 2, sec. 8; the constitution of Florida, art. 1, sec. 23; the constitution of Washington, art. 1, sec. 7.

18. U.S. Constitution, Amend. IV (1791): "The right of the people to be secure in their persons, houses, papers, and effects, against unreasonable searches and seizures, shall not be violated, and no warrants shall issue but upon probable cause, supported by oath or affirmation, and particularly describing the place to be searched, and the persons or things to be seized."

19. U.S. Constitution, Amend. XIV, Sec. 1 (1868): "nor shall any

the rights of individual autonomy, which the U.S. Supreme Court has recognized over the years.[20] It is in the latter sense—individual autonomy—that the term "right to privacy" is used in bio-ethics cases.[21]

2. Freedom from Unconsented Touching (Battery)

The second important distinction is between the right to privacy against governmental intrusion and the common-law tort principle that an individual has a right to be free from unconsented touching (battery) by any person. It is the tort concept of battery, rather than the constitutional right to individual autonomy, which is generally held to form the basis of the right of a competent person to refuse medical treatment.[22]

Although a thorough discussion of the common-law treatment of battery in the context of medical care is beyond the scope of this paper, it should be noted that the right to refuse treatment is not absolute, even for a competent adult. While it is generally recognized that "the right of a person to control his own body is a basic societal concept, long recognized in the common law,"[23]

State deprive any person of life, liberty, or property without due process of law."

20. See, for example, *Lochner v. New York*, 198 U.S. 45 (1905) (autonomy of contract in the employment setting, invalidating state laws regulating bakers' hours); *Meyer v. Nebraska*, 262 U.S. 390 (1923) (autonomy to make educational decisions, invalidating state laws that forbade teaching any language other than English); *Griswold v. Connecticut*, 381 U.S. 479 (1965) (autonomy of married persons to use contraceptives, invalidating state laws prohibiting the use of birth-control devices); *Roe v. Wade*, 410 U.S. 113 (1973) (autonomy to choose abortion, invalidating state laws protecting the life of the unborn child). But see *Bowers v. Hardwick*, 478 U.S. 186 (1986) (refusing to recognize the autonomy of homosexuals to engage in consensual sodomy, but reserving the decision with respect to sodomy by married couples).

21. Article 2, section 8 of the Arizona Constitution provides that "no person shall be disturbed in his private affairs, or his home invaded, without authority of law." In *Rasmussen v. Fleming*, 154 Ariz. 207, 741 P.2d at 682, the Arizona Supreme Court construed this language as providing for a right to refuse medical treatment.

22. See, for example, *Schloendorff v. Society of New York Hospital*, 211 N.Y. 125, 129-30, 105 N.E. 92, 93 (1914), quoted in *In re Conroy*, 98 N.J. 321, 346, 486 A.2d 1209, 1222 (1985).

23. *In re Conroy*, 98 N.J. at 346, 486 A.2d at 1221.

the cases also expressly recognize that the common-law right to refuse medical care is not absolute and must give way in the face of important governmental interests.[24] Recent cases have set out the countervailing interests as these:

(1) the preservation of life,
(2) the protection of interests of innocent third parties (e.g., children and other dependents),
(3) the prevention of suicide, and
(4) the maintenance of the ethical integrity of the medical profession.[25]

The main difference between the constitutional right to individual autonomy and the common-law right to refuse medical treatment is that the former is far broader. In a case asserting the common-law right to refuse treatment, the consent of the affected individual is the central issue. As a matter of common or judge-made law, standards governing consent can be modified in whole or in part by legislation. However, constitutional rights to privacy/autonomy operate as limits on the power of the state to make any regulation *at all* regarding the subject matter.[26] Decisions based squarely on constitutional concepts of individual autonomy reflect a judicial policy that seeks to remove the issue from legislative or executive competence altogether.[27]

24. *In re Conroy,* 98 N.J. at 348-54, 486 A.2d at 1223-26. See also *Jacobson v. Massachusetts,* 197 U.S. 11 (1905); *Application of President & Directors of Georgetown College, Inc.,* 331 F.2d 1000, 1008 (D.C. Cir. 1964), certiorari denied, 377 U.S. 978 (1964); *In re Caulk,* 125 N.H. 226, 480 A.2d 93, 96-97 (1984); *State v. Perricone,* 37 N.J. 463, 181 A.2d 751, certiorari denied, 371 U.S. 890 (1962).

25. See, for example, *Brophy v. New England Sinai Hospital, Inc.,* 398 Mass. 417, 497 N.E.2d at 634 (1986). See also *In re A.C.,* 533 A.2d 611, 539 A.2d 203 (D.C. 1987) (protection of others); *In re E.G., a Minor,* 161 Ill. App. 3d 75, 515 N.E.2d 285 (1987) (prevention of suicide: right of minor to reject blood transfusion for religious reasons).

26. Compare *In the Matter of Baby M,* 109 N.J. 396, 537 A.2d 1227 (1988), rejecting consent and an absolute right of procreational autonomy as theories of justification where important third-party interests are at stake.

27. See, for example, *Matter of Peter,* 108 N.J. 365, 529 A.2d 410, 427 (1987) ("Moreover, we find it difficult to conceive of a case in which the state could have an interest strong enough to subordinate a patient's right to choose not to be artificially sustained in a persistent vegetative state";

3. Causation, Intent, and Duty

The third major distinction requires brief mention of causation, intent, and duty.

Causation. Causation is perhaps the most difficult concept in tort law. To what extent it might be said that a given breach of duty (by act or omission) "causes" a given outcome is always debatable, and is immeasurably complicated by advances in medical technology that make it possible both to alter the natural progression of disease or disability and to prolong the dying process. Causation thus becomes an important consideration in death cases,[28] but not because it has anything to do with individual autonomy or the right to refuse medical treatment. Causation is important because it goes to the issue of culpability. If death resulted from natural causes, no one is "responsible." If human intervention was the cause, the death is either suicide or homicide.

Intent and Duty. The concepts of intent and duty are also useful in this inquiry—intent because it focuses on the state of mind of the actor, and duty because its reference point is a person *other than* the individual whose actions have been called into question, whether it be the person who desires death, a surrogate, or a physician or other health-care provider. Intent determines individual accountability under the criminal law and forms the basis of state culpability for violations of federal constitutional rights.

Duty plays the same role with respect to surrogate decision-making, whether the guardian is a family member, friend, or surrogate appointed by advance directive or court order. Whatever the source of the duty, the surrogate will be expected to act in the best interests of another and is accountable for all actions or failures to act that have the effect of harming those interests. Thus,

---

decided on non-constitutional grounds); *In re Colyer*, 99 Wash.2d 114, 119-21, 660 P.2d 738 (1983) (decided in part on constitutional grounds). For an interesting discussion of the relationship between the concepts of consent and constitutionally based privacy or autonomy, see *In the Matter of Baby M*, 109 N.J. 396, 468n.16, 537 A.2d 1227 (1988).

28. See, for example, *Matter of Peter*, 108 N.J. 365, 529 A.2d at 427-28 (specifically rejecting starvation and dehydration as the "cause" of death when nutrition and fluids are withheld; causation attributed to underlying condition).

whether the issue is one of intent or of duty, both intent and duty operate as limits on individual autonomy.

With respect to duty, one other point is noteworthy. Although the concept of "duty" as an aspect of American civil-rights discourse is not well developed, nowhere does it play a more important role than in the area of equal protection of the laws.

In most discussions relating to the nature of individual liberties, the central concern is whether the issue under scrutiny can be characterized as a matter of "right" (i.e., individual autonomy)—that is, shall the individual be at liberty to seek enjoyment of the right without governmental or private interference? This is the right of individual autonomy recognized under the rubric of the liberty protected by the Due Process Clause of the Fourteenth Amendment.

Yet careful reflection upon the concept of equal protection of the laws, which is protected by the very next clause of the Fourteenth Amendment,[29] discloses that while its *goal* is the equal enjoyment of the blessings of liberty, its *foundation* is an obligation to provide protection. To state that all persons are entitled to "equal protection of the laws" is merely to express a social duty that has been codified as a matter of constitutional law.[30]

29. U.S. Constitution, Amend. XIV, Sec. 1 (1868): "nor shall any State . . . deny to any person within its jurisdiction the equal protection of the laws."

30. Professor Alan Gewirth of the University of Chicago has described such duties as "strict" duties (i.e., those upon which the rights of others depend), and has contrasted them with "non-strict," or supererogatory, duties. See Gewirth, "Moral Foundations of Civil Rights Law," in *Symposium: The Religious Foundations of Civil Rights Law,* special issue of the *Journal of Law & Religion* 5 (1987): 157-58.

The duty to provide equal treatment has also been imposed by statute upon certain classes of individuals for the benefit of others who are deemed to be in need of special protection, including the elderly and the handicapped. See, for example, Title VII of the Civil Rights Act of 1964, 42 U.S.C. 2000e et seq. (prohibiting discrimination in employment on the basis of race, sex, national origin, and religion); Section 504 of the Rehabilitation Act of 1973, 29 U.S.C. 794 (1982) (prohibiting federally funded programs from discriminating against those with disabilities); Age Discrimination in Employment Act, 29 U.S.C. 621 (1982); 42 U.S.C. 1975c (a)(1) (1982) (giving the U.S. Commission on Civil Rights the jurisdiction to investigate discrimination on the basis of, among other things, "handicap"); The Education for All Handicapped Children Act of 1975,

Unlike the right to individual autonomy, the constitutional and legal right to equal protection of the laws requires that governments and individuals adapt their *own* behavior to a legally and socially acceptable standard imposed for the protection of others. A *constitutionally* based duty to protect thus arguably limits the power of both courts and legislatures to utilize concepts of individual autonomy to limit the protection afforded to those with disabilities.

C. The Role of Legal Fiction: Surrogate Decision-Making and Individual Autonomy

Recent cases place heavy reliance on proxy or surrogate decision-making regarding refusal of medical treatment. It is therefore necessary to make several further distinctions among cases:

(1) cases where the individual has clearly indicated his or her preference via an advance directive, such as a "living will" or durable power of attorney;

(2) cases where the indication of preference regarding medical treatment in the event the individual should someday become incompetent was made in passing in conversation with family or friends; and

(3) cases where the individual either made no statements at all, or was incapable of doing so (e.g., children and those with severe disabilities).

1. Individual Autonomy as an Act of Will

Only in cases where the individual has clearly indicated his or her preference via an advance directive, such as a "living will" or durable power of attorney, is it accurate to say that the individual's right to autonomy is exercised by proxy. Although this view has been criticized on the grounds that it permits past preferences to govern even though a person's desires may change over time,[31] there can be no question that, at least at the time the

---

20 U.S.C. 1400 (1982). See also Exec. Order No. 11,478, 34 Fed. Reg. 12,985 (1969) (prohibiting the federal government from discrimination in employment on the basis of, among other things, "handicap").

31. See *Rasmussen v. Fleming,* 154 Ariz. 207, 741 P.2d 674, 688n.24 (1987). See also R. Dresser, "Life, Death and Incompetent Patients: Con-

directive was given, the person now incompetent had chosen either a course of medical treatment or the individual he or she would entrust to make the choice about treatment.

Where the indication of preference regarding medical treatment in the event the individual should someday become incompetent was made in passing in the course of conversation with family or friends, the assertion that substituted judgment really effectuates the incompetent person's subjective intent or consent is more problematic.[32] In these cases the courts generally require "clear and convincing" evidence of the individual's intent, such as clearly articulated statements he or she made to family and friends.[33] But two courts have recently crossed the boundary into the third category by approving the decisions of close family members who refused nutrition and fluids on the incompetent relative's behalf based on what they *thought* he or she would have wanted.[34]

When there is no convincing evidence of the patient's desires, when the individual either has made no statements at all or is incapable of doing so (e.g., infants, young children, and those with severe disabilities), judicial assertions that the right to individual autonomy is the basis for third-party decision-making lie clearly within the realm of legal fiction.[35] Even though some courts feel that "by standards of logic, morality and medicine the terminally ill should be treated equally, whether competent or incompetent," and that no "societal policy objective is vindicated or furthered by treating the two groups of terminally ill differently,"[36] the fact remains that no amount of legal reasoning or justification will transmute a legal fiction into a factual reality. In

---

ceptual Infirmities and Hidden Values in the Law," *Arizona Law Review* 28 (1986): 373, 379-82.

32. In *In re Conroy*, 98 N.J. 321, 486 A.2d 1209 (1985), this inquiry is known as the "subjective" test.

33. See, for example, *In re Gardner*, 534 A.2d 947 (Me. 1987).

34. See *Matter of Jobes*, 108 N.J. 394, 529 A.2d 434, 446-47 (1987); *In re Guardianship of Grant*, 109 Wash.2d 545, 747 P.2d 445, 455-57 (1987).

35. *In re Conroy*, 98 N.J. 321, 486 A.2d 1209, 1246 (1985) (Handler, J., dissenting).

36. *Rasmussen v. Fleming*, 154 Ariz. 207, 741 P.2d 674 (1987), quoting *Eichner v. Dillon*, 73 A.D.2d 431, 464-65, 426 N.Y.S.2d 511, 542-43 (App. Div. 1980), modified under the name of *Matter of Storar*, 52 N.Y.2d, 420 N.E.2d 64 (1980), certiorari denied, *Storar v. Storar*, 454 U.S. 858 (1981) (italics omitted).

these cases, *someone else* makes the decision for the incompetent person or child, and the constitutional right to individual autonomy and the common-law right to refuse treatment become simply irrelevant. Justice Handler of the New Jersey Supreme Court has accurately captured the dilemma:

> The cases we presently consider dramatize the paradox in transporting the concepts of self-determination from competent to incompetent persons. . . . We, nevertheless, cling strongly to the belief that we can and should effectuate "self-determination" for the incompetent. Any determination, however, at best is only an optimistic approximation.[37]

2. A Guardian's Duty to Protect the Ward's "Best Interests"
Notwithstanding the rhetoric of individual autonomy that appears in these cases, the issue in most refusals of medical care is whether the proposed treatment is in the individual's best interests. Where competent adults are involved, the common law recognizes their right to refuse treatment even if that treatment is in their best interests. However, where the individual is incompetent because he or she is a child or is disabled, the issue is more properly considered as part of a discussion of the guardian's duty to act in the incompetent person's best interests, and of the law to see that those interests are protected notwithstanding the condition producing incompetency.

When an individual is not in the process of dying and his or her refusal of treatment, food, or fluids will result in certain death, the real question for the court is whether or not death is in the individual's best interests. Although phrased in terms of an analysis of "objective" factors, the process is in reality a determination of whether death "on balance" appears to be a net "benefit," given the person's condition.[38]

Where the individual is competent, the courts uniformly recognize his or her right to reject any treatment unless certain factors (see note 25) require a different result. Where the individual is severely disabled, a child, or otherwise incompetent,

---

37. *Matter of Jobes*, 108 N.J. 394, 529 A.2d at 454 (Handler, J., concurring).

38. See *Matter of Jobes*, 108 N.J. 394, 529 A.2d at 459n.14, 462 (Pollock, J., concurring), in which the phrasing is "the totality of the benefits and burdens of the patient's life in his or her present condition."

the same factors are relevant, but the individual is by definition incapable of exercising any right. Someone else must make the determination about treatment. If the intent is to avoid a clearly articulated judgment that death is in the incompetent person's "best interests," it advances neither the law nor citizens' respect for the legal system to hide behind such devices as the transparent fiction of individual autonomy, a definition of "treatment" that has now been expanded to include feeding and hydration,[39] and a narrow definition of "causation." But the incentive to do so is understandable. Once again Justice Handler has put his finger on the nub of the problem:

> The problems with the "best interests" analysis are straight-forward. In our society persons have different ideas about how the value of life is affected by the loss of brain function, the loss of cognitive ability, bodily deterioration, or unre-lievable extreme pain. A "best interests" standard assumes a consensus that is not there regarding when discontinua-tion of treatment is in a patient's best interests.[40]

In Handler's view, the solution is for the courts to "more directly confront, rather than finesse, the difficulties intrinsic to the objec-tive approaches" because the courts "must develop some variation of that [objective or "best interests"] approach to deal with the ex-treme cases where subjective approaches seeking individual self-determination are unavailing."[41] In other words, legal fiction can be stretched only so far. Eventually it will be impossible to avoid the ultimate issue the courts so strenuously seek to avoid: whether or not a severely handicapped individual is "better off" dead.[42] It is the ultimate eugenic quandary: whether life with certain disabilities is preferable to death or, as some put it, to "non-existence."[43]

39. This is not to say, however, that treatment, nutrition, or hydra-tion must be provided in every case. Where there are sound medical reasons for not providing them or for withdrawing them, the decision is based on medical factors, not a decision that death is preferable to the individual's current condition.

40. *Matter of Jobes*, 108 N.J. 394, 529 A.2d at 457-58 (Handler, J., con-curring).

41. *Matter of Jobes*, 108 N.J. 394, 529 A.2d at 458 and 458n.12.

42. See *Matter of Jobes*, 108 N.J. 394, 529 A.2d at 458n.12 (implicitly criticizing such avoidance).

43. In *Becker v. Schwartz*, 46 N.Y.2d 401, 411-12, 413 N.Y.S.2d 895,

By increments, the courts rationalize the use of eugenic factors to determine which lives are worth living,[44] and the law reaches the point where, as Reverend Richard John Neuhaus has succinctly put it, "human rights are coterminous with the individual's ability to claim and exercise [them]."[45] It is at this point that the conflict between rights of privacy and individual autonomy recognized under the Due Process Clause and the duty imposed by the Equal Protection Clause becomes apparent: by its very terms, the concept of equal protection of the laws imposes a duty upon government to *protect* all persons equally.

D. The Role of Legal Process and Precedent

Precedent is a critical factor in the judicial development of the law. The rule of *stare decisis* (literally, "the matter has been decided") exerts a strong influence on both common law and constitutional decision-making, giving both structure and predictability to judge-made law. The difficulty, however, is that policy choices, once made, are extraordinarily difficult to overturn, even in the face of strong arguments that the decisions in the initial cases on which they were based are wrong.[46]

900-01, 386 N.E.2d 807, 812 (1978), Judge Jasen formulated the quandary as follows:

> Whether it is better never to have been born at all than to have been born with even gross deficiencies is a mystery more properly to be left to the philosophers and the theologians. Surely the law can assert no competence to resolve the issue, particularly in view of the very nearly uniform high value which the law and mankind has placed on human life, rather than its absence.

44. See *United States v. Altsoetter* (Military Tribunal III, 1947) ("The Justice Case"; eugenics courts). Compare *Skinner v. Oklahoma*, 316 U.S. 535 (1942) (sterilization of certain criminals); *Buck v. Bell*, 274 U.S. 200 (1927) (sterilization of the mentally incompetent).

45. Neuhaus, "Nihilism without the Abyss: Law, Rights, and Transcendent Good," in *Symposium: The Religious Foundations of Civil Rights Law*, special issue of the *Journal of Law & Religion* 5 (1987): 53, 57 (referring to *Roe v. Wade*, 410 U.S. 113 [1973]). Compare *Bowen v. American Hosp. Ass'n*, 410 U.S. 610 (1986).

46. Compare *Matter of Jobes*, 108 N.J. 394, 529 A.2d at 461 (Handler, J., concurring): "Because of our incertitude, we cannot say that particular right-to-die decisions were clearly correctly decided while other deci-

This is especially true in bio-ethics cases. Whether the cases involve abortion, euthanasia, in vitro reproduction, or organ "harvesting" from anencephalic children for transplantation into others, most can be classified as "hard cases," pitting excruciatingly difficult human dilemmas against the requirements of law. Although the courts are not well-equipped to deal with such issues, the grim reality is that the task of deciding who shall live and who shall die falls upon lawyers and judges, whose only distinction is their legal training and judicial office.[47] Given the desperate circumstances of most of the disabled individuals involved and the emotional and financial burdens of their families, death invariably appears to be the only "reasonable" solution.[48]

Because the arguments favoring a "quality-of-life" approach derive their persuasive force from the facts of individual cases rather than from a logical extension of long-established and accepted legal principles, a "case-by-case" approach to decision-making which accepts as its implicit starting point the proposition that death is in the "best interests" of those who must live with certain disabilities[49] will inevitably erode the protection existing law affords to disabled people.[50] The difficulty lies not in allowing nature to take its course whenever continuation of treatment would

---

sions were dangerously wrong. Decisions of such painful difficulty cannot be so easily rejected or so quickly applauded." See also *City of Akron v. Akron Center for Reproductive Health*, 462 U.S. 416, 419n.1 (1983) (opinion of the Court per Powell, J., discussing the role of *stare decisis* in constitutional law in the face of an argument that the court's approach to the issue was wrong as a matter of constitutional law).

47. See *Matter of Jobes*, 108 N.J. 394, 529 A.2d at 460-61 (Handler, J., concurring) (recognizing that the inquiry "has as much to do with judicial attitudes as with judicial decisions," and describing the process "as a form of judicial deregulation reflecting a deference to individual autonomy and to the professional relationship in which treatment decisions are made").

48. See *Matter of Jobes*, 108 N.J. 394, 529 A.2d at 460 ("to perpetuate life in only its most primitive form . . . offends fundamental sensibilities, and it should stop").

49. This assumption is implicit whenever it is urged that "decisionmakers should consider the totality of the benefits and burdens of the patient's life in his or her present condition" (*Matter of Jobes*, 108 N.J. 394, 529 A.2d at 456 [Handler, J.]; 462 [Pollock, J., concurring]). Where the balance tips against life, death apparently is in the patient's best interests.

50. See note 30.

be medically contraindicated, counterproductive, or otherwise inhumane, but in the assumption that policymakers of any sort are competent to judge the quality of another person's life.

It is the arrogation of such competence that constitutes the critical exception to the natural-rights principles upon which current law is based. Once established in precedent, the legitimacy of the reasoning becomes the philosophical underpinning for its extension. Needless to say, any attempt to re-impose the pre-existing rules or to limit the advance of the newly established legal principle is seen as a "step backward" into an ethically or legally less desirable formulation of public policy.

Analysis of recent cases involving handicapped newborns and incompetent adults shows that the gradual acceptance of acts intended to cause death in situations where a life appears, on balance, not to be worth preserving follows precisely this pattern.[51] Whereas the law has only recently evidenced concern for the claims of the disabled, the constitutional and common-law analysis employed by the courts in "right to die" cases is one which is clearly sympathetic to the application of the eugenic ideas that disability law is designed to eliminate, but it is not brave enough to state simply that the incompetent subject is understood as not fit to live.[52] Thus advocates for a quality-of-life approach

51. One of the best examples of the incremental nature of the process is its development in New Jersey, which can be traced through the following cases: *In re Karen Quinlan*, 70 N.J. 10, 355 A.2d 647 (1976) (withdrawal of respirator from comatose patient); *Procanik v. Cillo*, 97 N.J. 339, 478 A.2d 755 (1984) (accepting concept of wrongful life); *In re Conroy*, 98 N.J. 321, 486 A.2d 1209 (1985) (withholding food from non-comatose patient in waking stupor); *Matter of Peter*, 108 N.J. 365, 529 A.2d 410 (1987) (withholding food from comatose patient); *Matter of Jobes*, 108 N.J. 394, 529 A.2d 434 (1987) (withholding food from patient with severe brain damage); *Matter of Farrell*, 108 N.J. 335, 529 A.2d 404 (1987) (permitting withdrawal of respirator from competent individual desiring to die).
52. In *Matter of Jobes*, 108 N.J. 394, 529 A.2d at 445, the New Jersey Supreme Court appears to take a swipe at "[those] who [would not] treat the patient as a person, [but as] a symbol of a cause." To the extent that any individual would demand measures which either do no good or harm the patient in a futile attempt to preserve life, the court's criticism is well-taken. But to the extent that the criticism can be taken as a slap at those who object to the court's reasoning, it is on thin ice indeed: resistance to the eugenic mind-set should, in this writer's view, be a cause célèbre.

content themselves with legal arguments such as individual autonomy that on the surface have little to do with the eugenic ethic, and the courts, as yet unprepared to accept eugenic reasoning directly, decide them on those grounds.

A chronological review of the cases produces an extraordinarily clear illustration of this principle. *In re Karen Quinlan*[53] rests in part on concepts of individual autonomy, personhood "in the whole sense," and "meaningful life" accepted by the U.S. Supreme Court in the abortion cases,[54] and it was the first major case to apply that reasoning in the context of medical treatment for incompetent adults. The right to privacy—in the sense of personal autonomy—thus became the justification for "letting nature take its course" with respect to the ordinary treatment of incompetent disabled persons. That the person who will die is incompetent to make the choice has presented no obstacle: the theory of "substituted judgment" provides a convenient legal ruse to mask the reality that the incompetent person's death occurs as a result of an act calculated by a guardian to cause the death of his or her ward.[55]

Since *Roe v. Wade* (1973) and *In re Karen Quinlan* (1976), the courts have approved the non-treatment and resulting starvation death of a handicapped newborn,[56] the withholding of medically indicated treatment from a handicapped infant,[57] the denial of medically indicated surgery to a twelve-year-old boy with Down's

53. *In re Karen Quinlan,* 70 N.J. 10, 355 A.2d 647 (1976).

54. See *Roe v. Wade,* 410 U.S. 113 (1973); *Doe v. Bolton,* 410 U.S. 179 (1973).

55. Dr. Gregory Pence has noted that "Dutch medicine had long before rejected the American view that actively killing a terminal patient differs ethically from withdrawing care." See his article entitled "Do Not Go Slowly into that Dark Night: Mercy Killing in Holland," *American Journal of Medicine* 84 (1988): 139, 140.

56. See *In re Infant Doe,* No. GU8204-00 (Cir. Ct., Monroe County, Ind., 12 Apr. 1982), affirmed under the name of *State of Indiana on Relation of Infant Doe by Guardian,* No. 482 S.139 (Ind. S. Ct., 27 May 1982), No. 482 S.140 (Ind. S. Ct., 26 Apr. 1983), certiorari denied, 464 U.S. 961 (1983).

57. See *United States v. University Hospital,* 729 F.2d144 (2d Cir. 1984), affirmed, *Bowen v. American Hosp. Ass'n,* 410 U.S. 610 (1986) (Baby Jane Doe). See also *Weber v. Stony Brook Hosp.,* 60 N.Y.2d 208, 469 N.Y.S.2d 63, 456 N.E.2d 1186 (1983), affirmed, 95 A.D.2d 587, 467 N.Y.S.2d 685 (App. Div., 2d Dept., 1983).

syndrome,[58] and the starvation and dehydration deaths of numerous incompetent adults described as "vegetative."[59] The common law has also witnessed the creation of two new civil torts: "wrongful birth," the essence of which is the duty of a physician to warn of potential birth defects in time for a eugenic abortion,[60] and "wrongful life," an action brought *on behalf of a child born with handicaps* for the harm allegedly caused by a life which is alleged to be worse than not existing.[61] Two recent cases have gone even farther:

58. See *In re Phillip B., A Minor*, No. 66103 (Super. Ct., Santa Clara County, Cal., 27 Apr. 1978), affirmed, 92 Cal. App. 3d 796, 156 Cal. Rptr. 48 (1979), certiorari denied, 445 U.S. 749 (1980).

59. The term "vegetative state" has been described as follows: "Vegetative state describes a body which is functioning entirely in terms of its internal controls. It maintains temperature. It maintains heartbeat and pulmonary ventilation. It maintains digestive activity. It maintains reflex activity of muscles and nerves for low level conditioned responses. But there is no behavioral evidence of either self-awareness or awareness of the surroundings in a learned manner." See *Matter of Jobes*, 108 N.J. 394, 529 A.2d 434, 438 (1987).

The same state was described as follows by the President's Commission for the Study of Ethical Problems in Medicine and Biomedical and Behavioral Research: "Personality, memory, purposive action, social interaction, sentience, thought, and even emotional states are gone. Only vegetative functions and reflexes persist. If food is supplied, the digestive system functions, and uncontrolled evacuation occurs; the kidneys produce urine; the heart, lungs, and blood vessels continue to move air and blood; and nutrients are distributed to the body" (*Deciding to Forego Life-Sustaining Treatment*, pp. 174-75).

60. See, for example, *Robak v. United States*, 658 F.2d 471 (7th Cir. 1981) (Alabama law); *Turpin v. Sortini*, 31 Cal.3d 220, 182 Cal. Rptr. 337, 643 P.2d 954 (1982); *Fassoulas v. Ramey*, 450 So.2d 822 (Fla. 1984); *Blake v. Cruz*, 108 Idaho 253, 698 P.2d 315 (1984); *Berman v. Allan*, 80 N.J. 421, 404 A.2d 8 (1979); *Speck v. Finegold*, 497 Pa. 76, 439 A.2d 110 (1981); *Dumer v. St. Michael's Hosp.*, 69 Wis.2d 766, 233 N.W.2d 372 (1975).

61. See *Curlender v. Bio-Science Laboratories*, 106 Cal. App. 3d 811, 165 Cal. Rptr. 477 (1980); *Continental Casualty Co. v. Empire Casualty Co.*, 713 P.2d 384 (Colo. App. 1985), cert. granted (Colo. Sup. Ct., 13 Jan. 1986); *Procanik v. Cillo*, 97 N.J. 339, 478 A.2d 755 (1984); *Harbeson v. Parke-Davis, Inc.*, 98 Wash.2d 460, 656 P.2d 483 (1983). It is important to note, however, that the recovery permitted in these cases is limited to the costs of extraordinary care, and does not extend to the alleged "damage" caused by life itself. Every other jurisdiction which has considered the issue has held that no such cause of action exists. See, for example, *Elliott v. Brown*, 361 So.2d 546 (Ala. 1978); *Turpin v. Sortini*, 31 Cal.3d 220, 182 Cal. Rptr. 337, 643 P.2d 954 (1982); *Siemiemiec v. Lutheran Gen. Hosp.*, 117 Ill.2d 230, 512

one reportedly involved a request for a lethal injection by a severely disabled, competent patient who died before the court could rule; the other raised euthanasia as a defense in response to a charge of first-degree murder.

Thus Peter Singer was undoubtedly correct when he stated in 1983 that "the ethical outlook that holds human life to be sacrosanct—I shall call it the 'sanctity-of-life view'—is under attack," and that "the first major blow . . . was the spreading acceptance of abortion."[62] The issue now is what the law should do about it.

Given the validity of Singer's empirical observations concerning the direction of the law, it is critically important to distinguish between the demands of existing law, which prefers life,[63] and the concerns of ethics (or ethicists), which speak to "what ought to be" as a matter of correct behavior. Established laws are not the equivalent of moral principles or ethical norms, although it is possible to make moral and ethical arguments to support or attack them. Established laws are simply legal rules that bind both courts and citizenry unless and until changed through legally permissible means. Ethical or moral arguments to the effect that given legal rules should be changed to reflect a normative judgment other than that reflected in current law will not suffice to justify judicial attempts to change or avoid existing policy. In a representative democracy, the all-important legal questions—"changed how, and by whom?"—are too important to avoid, for it is only in the legislative arena that the difficulty of "hard cases" can be moderated through discussion of the duty of those charged with the care of those who cannot care for them-

---

N.E.2d 691 (1987), reversed 134 Ill. App. 3d 823, 480 N.E.2d 1227 (1985); *Becker v. Schwartz*, 46 N.Y.2d 401, 413 N.Y.S.2d 895, 386 N.E.2d 807 (1978); *Dumer v. St. Michael's Hosp.*, 69 Wis.2d 766, 233 N.W.2d 372 (1975); *Jacobs v. Theimer*, 519 S.W.2d 846 (Tex. 1975). Curlender's "wrongful life" theory was disapproved by the California Supreme Court in *Turpin v. Sortini*. Curlender's holding that parents might be held liable was negated by amendment of the California Civil Code. See Cal. Civ. Code, Sec. 43.6 (West 1988).

62. Singer, "Sanctity of Life or Quality of Life?" p. 128.

63. See *Azzolino v. Dingfelder*, 315 N.C. 103, 337 S.E.2d 528 (1985), quoting *Becker v. Schwartz*, 46 N.Y.2d 401, 411-12, 413 N.Y.S.2d 895, 900-01, 386 N.E.2d 807, 812 (1978).

selves. Only in this manner is it possible to avoid the tendency of hard cases to make bad law.

II. WHOSE "RIGHT TO A GOOD LIFE"? THE FUNCTIONAL DEFINITION OF THE HUMAN "PERSON"

A. The Interaction of Legal Reasoning and Bio-ethical Opinion

The need to resolve bio-ethical/legal controversies expeditiously often diverts attention from the significant incremental impact that the process of legal reasoning has on the vitality of basic legal principles based on natural-rights theory. It is not particularly important whether the process is described as a "slippery slope" or in some other fashion that conveys a gradual disintegration of a concept or value once clear in the law. It is the *shift in attitude* which is critical.

Shifts in legal and judicial attitudes toward the protection of vulnerable children and adults do not, however, occur in a vacuum. The writings of medical experts and bio-ethicists have had a substantial effect on the development of case law. Not only do such writings provide background research materials for the courts' general information, but the factual records of the cases themselves often include testimony by experts in medicine, bio-ethics, philosophy, and morals.[64] And once a medical, moral, or philosophical point has been accepted by a significant segment of

---

64. For example, in *Delio v. Westchester County Medical Center*, 516 N.Y.S.2d 677, 683 (Å.D. 2 Dept., 1987), the New York courts rested their holding in part on the testimony of a Jesuit priest and philosophy professor at Fordham University, who testified concerning the Vatican's position on life-sustaining measures. In *In re Guardianship of Grant*, 109 Wash.2d 545, 747 P.2d 445, 454n.2 (1987), the Washington Supreme Court relied on the same testimony, now even further removed from the possibility of cross-examination, to support its reliance on Delio's holding that food and water "supplied by artificial means should be evaluated in the same manner as any other medical procedure." See also *In re Estate of Prange*, 166 Ill. App. 3d 1091, 520 N.E.2d 946 (1988), vacated without opinion, 121 Ill.2d 570, 1988 WL 5111 (19 May 1988) (slip copy) (testimony by experts in medical ethics and by a priest and a nun concerning patient's wishes).

experts, the courts apparently feel free to utilize such opinions as precedent for their own holdings concerning the medical, moral, or philosophical points involved in the case. From that point, it is merely a matter of time before "expert" opinion becomes controlling in its own right.[65]

In 1970, an editorial in *California Medicine,* the journal of the California Medical Association, suggested that the time was right for the adoption of a "new ethic" which would take a more relativistic approach toward the valuation of individual human lives.[66] It accurately noted that "medicine's role with respect to changing attitudes toward abortion will be a prototype of what is to occur," and went on to predict that "one may anticipate further developments of these roles as problems of birth control and birth selection are extended, inevitably to death selection and death control, whether by the individual or by society, and further public and professional determination of when and when not to use scarce resources."[67] Although the courts have yet to adopt such reasoning explicitly, a considerable number of experts have done so, and their reasoning has had considerable influence on the development of the law.[68]

B. Defining the Legal Issues

The place to begin an examination of the impact of shifting bioethical opinions on who should be protected by the law as a legal

65. The prior note points out that *In re Guardianship of Grant* rests in part on the holding in Delio. Delio in turn was based on the reasoning of *In re Conroy,* 98 N.J. 321, 486 A.2d 1209 (1985), and *Brophy v. New England Sinai Hospital, Inc.,* 398 Mass. 417, 497 N.E.2d 626 (1986). The point here is that great reliance is put on cases involving similar facts, and that harder cases are judged by the same initial premises.

66. "A New Ethic for Medicine and Society," editorial, *California Medicine,* Sept. 1970, p. 68.

67. "A New Ethic for Medicine and Society," p. 68.

68. See, for example, *Matter of Storar,* 52 N.Y.2d at 373n.3, 420 N.E.2d at 75-76n.3 (Jones, J., dissenting in part, and noting that recent surveys suggest that the majority of practicing physicians now approve of passive euthanasia and believe that it is practiced by members of the profession); American Medical Assn., Council on Ethical and Judicial Affairs, "Statement on Withholding or Withdrawing Life Prolonging Medical Treatment," 15 Mar. 1986.

"person" is section one of the Fourteenth Amendment to the Constitution of the United States. The Due Process Clause prohibits government action designed to limit or otherwise restrict the rights of "any person" to "life, liberty, or property," and the Equal Protection Clause prohibits denials of "equal protection of the laws."[69] The operative legal terms are "any person," "life," "liberty," and "protection." The most effective means for gradually eroding the natural-rights ethic protected by the original meaning of these terms is to ignore it and to replace it with legal devices that rationalize decisions reflecting a preference for death. The operative terms remain the same, but their substance reflects the "new" ethic rather than the original one. Only after the new ethic is well established in precedent does explicit acceptance of quality-of-life reasoning appear.[70]

Quality-of-life arguments generally turn on one of two possible legal approaches: judicial expansion of the "liberty" (autonomy/self-determination) interest protected by the Due Process Clause and the common law, and limitation of the class of persons covered by presently existing law governing the right of the disabled and handicapped to equal treatment.[71] Except in rare cases where the issue of legal "personhood" is central to the controversy—cases such as abortion and perhaps future cases involving the degree of legal protection to be afforded children conceived and brought to term without maternal assistance[72]—the

69. U.S. Constitution, Amend. XIV, Sec. 1 (1868): "nor shall any State deprive any person of life, liberty, or property without due process of law, nor deny to any person within its jurisdiction the equal protection of the laws."

70. See *Matter of Jobes*, 108 N.J. 394, 529 A.2d at 459n.12 (Handler, J., concurring) (pointing out the degree to which the New Jersey Supreme Court has sought to avoid the charge that it is employing quality-of-life reasoning).

71. See, for example, *Bowen v. American Hosp. Ass'n*, 410 U.S. 610 (1986) (restrictive interpretation of existing law); *American Academy of Pediatrics v. Heckler*, 561 F. Supp. 395 (D.D.C. 1983) (same, with additional reliance on "privacy" concerns); *In re Conroy*, 98 N.J. 321, 486 A.2d 1209 (1985) (common-law self-determination, power of the state as *parens patriae* for incompetents).

72. See S. Elias & G. J. Annas, "Social Policy Considerations in Noncoital Reproduction," *Journal of the American Medical Association* 255 (1986): 62-68; G. G. Blumberg, "Legal Issues in Nonsurgical Human

courts generally need not directly address the question "Who is a person under law?" It is possible to address the related question of "What duty does the law impose to protect vulnerable individuals?" without appearing to have an opinion on the question of personhood at all.

Whether the matter is approached directly or indirectly, however, the foreseeable and intended result of reasoning based on a quality-of-life perspective is the same: the class of "persons" protected by the law is defined functionally—by what an individual can *do* or *feel*—rather than by reference to their nature— that is, by what they *are*.

In Part I of this paper I pointed out that when the rationale is nonconstitutional or a mixture of common-law and constitutional analysis, as is often the case where an incompetent adult is the subject of the proceedings, reliance is placed upon a form of individual liberty that is said to be exercised by proxy.[73] In cases where the subject is a seriously ill or handicapped child or infant, however, a rationale of individual liberty or "self-determination" would be far too transparent a fiction for most courts. In these cases, a quality-of-life ethic—"the best interests of the child"—is simply infused into the traditional standard in juvenile medical-neglect cases,[74] and the means for implementing it is virtually identical to that employed in the case of abortion: a liberty-based "privacy" right of the parents to make whatever medical decisions they consider to be in the "best interests" of themselves and

---

Ovum Transfer," *Journal of the American Medical Association* 251 (1984): 1178-81; H. O. Tiefel, "Human in Vitro Fertilization: A Conservative View," *Journal of the American Medical Association* 247 (1982): 3235-42. Compare *In re Baby M*, 109 N.J. 396, 537 A.2d 1227 (1988) (surrogacy), affirmed in part and reversed in part, 217 N.J. Super. 313, 525 A.2d 1128 (1987); *Sherwyn v. Dept. of Social Services*, 173 Cal. App. 3d 52, 218 Cal. Rptr. 778 (2d Dist. 1985) (challenge to legislation limiting surrogacy dismissed on standing grounds).

73. See sections B and C of Part I.

74. See, for example, Cal. Penal Code, Sec. 11165 (West 1980); Colo. Rev. Stat., Sec. 18-6-401 (1978); D.C. Code Ann., Sec. 6-2101 (1981); Minn. Stat. Ann., Sec. 626.556 (West 1982); N.Y. Soc. Serv. Law, Sec. 412 (McKinney 1983). See the text accompanying notes 40-45 for a discussion of the problems with the "best interests" test as applied to incompetent patients generally.

their child.[75] And this is so even where the decision is based on an explicit statement that the child "would be better off dead."[76]

Under current case law, a handicapped infant will not be entitled to legal protection unless and until his or her parents accept him or her as fully human or the state demonstrates that the child will be functional.[77] Until that time, the child has no rights of his or her own which are independently enforceable by the state, not even those which are expressly designed to protect his or her rights as a handicapped person.[78]

75. Examination of the cases demonstrates that the courts will not hesitate to intervene in family affairs whenever they determine that a child is in need of treatment, even in the face of a parental claim that they have a constitutional right to refuse treatment. Since the issue is framed as one involving alleged parental neglect, sincere parental concern and affection for the child are not controlling factors. Some of the major factors to be considered are the effect of non-treatment either on the child's chance for life or on his/her emotional or psychological well-being; the benefits to be derived from the proposed treatment; the risks and potential side effects of the proposed treatment; medical testimony about the propriety of the proposed course of action; the child's preference where available; the effect of delaying the implementation of a proposed course of action; and whether or not an emergency exists or whether the child's life is in imminent danger. Recent case law indicates that in certain instances the parents may have no rights at all if the courts deem it necessary to exclude them altogether from the decision-making process. It is, therefore, rather anomalous for courts and commentators to rely on an asserted parental right to self-determination when the issue is their right to refuse medical treatment for their handicapped child. To do so is to confuse what the law is (i.e., courts will intervene as a matter of course) with what such advocates think it should be (i.e., that parents should have unfettered discretion when the child is disabled). Case references for the propositions cited in this note can be found in R. Destro, "Quality of Life Ethics and Constitutional Jurisprudence: The Demise of Natural Rights and Equal Protection for the Disabled and Incompetent," p. 98n.114.

76. R. Destro and W. Moeller, "Necessary Care for the Retarded Child: The Case of Phillip Becker," *Human Life Review* 4 (1980): 81, quoting Transcript of Record, *In re Phillip B., A Minor*, No. 66103 (Super. Ct., Santa Clara County, Cal., 27 Apr. 1978), affirmed, 92 Cal. App. 3d 796, 156 Cal. Rptr. 48 (1979), certiorari denied, 445 U.S. 749 (1980).

77. Compare *Roe v. Wade*, 410 U.S. 113 (1973).

78. See *Bowen v. American Hosp. Ass'n*, 410 U.S. 610 (1986) (Sec. 504 of the Rehabilitation Act of 1973).

In the case of adults with severe disabilities, the assumption appears to be that such individuals have lost the functional qualities which make them "human" and their lives worth living. The critical point for the disabled adult is that the person is not dead but in a state—sometimes but not always described as "vegetative"—such that it appears to the individual or to the outside observer that life isn't worth living anymore. The legal issue in these cases is not simply whether the individual should be permitted to die as a result of disease or disability, but whether the existence or extent of disease or disability is sufficient to justify steps taken with the intent to cause death—steps, such as the withdrawal of food or fluids, which would not be tolerated in cases where the individual was not seriously disabled. If the disabled person, or a "reasonable person" identifying with that situation, would prefer death to his or her present state of health or incompetence, it is argued that society has no further duty to protect him or her from legal harm,[79] and that his or her "right to die" (or to refuse treatment) should take precedence.

C. From Right to Duty: The Law as Guarantor of a Good Life

The authors of an article which appeared in the *Archives of Internal Medicine* in 1985 argued that the growing acceptance of the practice of withholding fluids from patients is troublesome "because it may bear the seeds of unacceptable social consequences." They went on to explain,

> We have witnessed too much history to disregard how easily a society may disvalue the lives of the "unproductive." The "angel of mercy" can become the fanatic, bringing the "comfort" of death to some who do not clearly want it, then to others who "would really be better off dead," and finally, to classes of "undesirable persons," which might include the terminally ill, the permanently unconscious, the severely senile, the pleasantly senile, the retarded, the incurably or chronically ill, and perhaps, the aged. . . . In the current environment, it may well prove convenient—and

79. President's Commission for the Study of Ethical Problems in Medicine and Biomedical and Behavioral Research, *Deciding to Forego Life-Sustaining Treatment*, p. 136.

all too easy—to move from recognition of an individual's "right to die" (to us, an unfortunate phrasing in the first instance) to a climate enforcing a "duty to die."[80]

Certain policymakers apparently have the same idea. In March 1984, Richard D. Lamm, at that time governor of Colorado, shocked a group of elderly listeners with the statement that elderly people with terminal illnesses "have a duty to die and get out of the way" because the cost of treating them with the new technologies ruins the nation's economic health and hampers the ability of younger people "to build a reasonable life."[81] Although Lamm later attempted to "clarify" his statement, the essential point had been made: the elderly are important only insofar as they accommodate the needs of the young. Those "who die without having life artificially extended are similar to 'leaves falling off a tree and forming humus for the other plants to grow up.'"[82]

That the promise of youth and vitality and the cost of age and disability are the focus of current legal policy is readily apparent. The language of the cases leaves no doubt. When the legitimacy of providing a proposed medical treatment for an elderly person or a disabled person of any age is analyzed in terms of the probability that it will restore a "normal, healthy life,"[83] the assumption is that the goal of medical treatment is to restore the individual to that state. Further, cases involving the elderly and those with disabilities note that such individuals are afflicted with "infirmities that limit their independence and deprive them of dignity."[84] A cost-benefit analysis is then applied to determine whether nature should simply be allowed to take its course.[85] In medical-neglect cases where the child has a disability, different

80. M. Siegler and A. Weisbard, "Against the Emerging Stream: Should Fluids and Nutritional Support Be Discontinued?" *Archives of Internal Medicine* 145 (1985): 129, 130-31.

81. Lamm, cited in the *New York Times*, 29 Mar. 1984, p. A16, col 5.

82. *Ibid.*

83. B. Mishkin, "A Matter of Choice: Planning Ahead for Health Care Decisions," an information paper prepared for use by the Special Committee on Aging, U.S. Senate (Wasington: GPO, 1986).

84. *Ibid.*

85. One court has summarized the factors to be considered as follows:

    1. the age of the patient

standards of child abuse are applied,[86] and the cases unabashedly use a form of cost-benefit analysis as the rationale.[87]

One recent case involved a seventy-four-year-old nursing-

---

2. the life expectancy with or without the procedure contemplated
3. the degree of present and future pain or suffering with or without the procedure
4. the extent of the patient's physical and mental disability and degree of helplessness
5. statements, if any, made by the patient which directly or impliedly manifest his views on life prolonging measures
6. the quality of the patient's life with or without the procedure, i.e., the extent, if any, of pleasure, emotional enjoyment or intellectual satisfaction that the patient will obtain from prolonged life
7. the risks to life from the procedure contemplated as well as its adverse side effects and degree of invasiveness
8. religious or ethical beliefs of the patient
9. views of those close to him
10. views of the physician
11. the type of care which will be required if life is prolonged as contrasted with what will be actually available to him
12. whether there are any overriding State *parens patriae* interests in sustaining life (e.g., preventing suicide, integrity of the medical profession or protection of innocent third parties, such as children).

Matter of Beth Israel Medical Center, 519 N.Y.S.2d 511, 517 (Supp. 1987).

86. See, for example, *In re Phillip B., A Minor*, No. 66103 (Super. Ct., Santa Clara County, Cal., 27 Apr. 1978), affirmed, 92 Cal. App. 3d 796, 156 Cal. Rptr. 48 (1979), certiorari denied, 445 U.S. 749 (1980). After the U.S. Supreme Court denied certiorari in *Phillip B.*, the Superior Court of Santa Clara County, California, transferred guardianship of Phillip Becker to the couple who sought to have the life-saving surgery performed over the objections of Phillip's parents. See *Guardianship of Phillip Becker*, No. 101981 (Super. Ct., Santa Clara County, Cal., 7 Aug. 1981). For a detailed discussion of the Becker case, see Destro and Moeller, "Necessary Care for the Retarded Child: The Case of Phillip Becker." See also testimony of Adrienne Asch and Mary Jane Owen, U.S. Commission on Civil Rights, Transcript of Hearing on the Protection of Handicapped Newborns, vol. 3, 26-27 June 1986, at 120-51. Compare *Bowen v. American Hosp. Ass'n*, 410 U.S. 610 (1986). But see Child Abuse Prevention and Treatment Act, 42 U.S.C.A. Sec. 5102(3) (West Supp. 1985).

87. See, for example, Shaw, "Defining the Quality of Life," *Hastings Center Report* 7 (1977): 11. In this article Shaw suggests the use of a for-

home resident, partially paralyzed and aphasic, who had suffered a stroke: the question was whether or not an amputation should be performed to prevent gangrene from taking its course and killing her.[88] While the court appeared to be persuaded by the physician's contention that the surgery would be both medically and ethically unsound, given the mutilation and risk that would be involved because of her frail condition, its rationale was considerably broader: "no life-shortening course of action should be contemplated unless the patient is, at the very least, suffering from severe and permanent mental and physical debilitation and with a very limited life expectancy."[89] Its definition of those terms in light of the facts of the case is revealing:

> [She] is presently without any functioning intellect or meaningful awareness of the life around her. Though not entirely comatose, *she is, for all practical purposes merely existing.* She cannot understand or speak. Her paralysis has left her immobile on her right side and her left side is now affected. Thus, she will never be able to move on her own. . . .
>
> Life has no meaning for her. She derives no physical or emotional pleasure of any degree, nor any intellectual satisfaction in her day to day existence. She will remain completely dependent on others. . . .
>
> In sum, other than to afford this patient the potential for limited additional life, the proposed surgery offers her no other benefit. If performed it would not to any degree return this patient to an integrated functioning or cognitive existence.[90]

---

mula, QL = NE × (H + S) to make treatment decisions where QL = Quality of Life; NE = Natural Endowment, both physical and intellectual; H = the contribution of Home and family; and S = the contribution of Society). See also Gross, Cox, Tatyrek, Pollay, and Barnes, "Early Treatment and Decision Making for the Treatment of Myelomeningocele," *Pediatrics* 72 (1983): 450; U.S. Commission on Civil Rights, Transcript of Hearing on the Protection of Handicapped Newborns, vol. 2, 26-27 June 1986.

88. *Matter of Beth Israel Medical Center*, 519 N.Y.S.2d 511 (Supp. 1987).

89. *Matter of Beth Israel Medical Center*, 519 N.Y.S.2d at 517-518 (Supp. 1987).

90. *Matter of Beth Israel Medical Center*, 519 N.Y.S.2d at 517-518, emphasis added.

The explicit assumption in this and other such cases is that the ultimate aim of medicine is to cure or rehabilitate, an assumption which lies at the very core of the quality-of-life ethic. In 1949, Leo Alexander, an observer at the trials of the Nazi war criminals at Nuremberg, wrote that what allowed the terror to begin was a shift in the medical profession's view of the purpose of medicine. Killing became, in the words of Robert J. Lifton's recent work, "medicalized,"[91] and the law simply took advantage of an already fertile climate of eugenic opinion. Alexander wrote,

> The original concept of medicine and nursing was not based on any rational or feasible likelihood that they could actually cure and restore but rather on an essentially maternal or religious idea. The Good Samaritan had no thought of nor did he actually care whether he could restore working capacity. He was merely motivated by the compassion in alleviating suffering. [J.] Bernal states that prior to the advent of scientific medicine, the physician's main function was to give hope to the patient and to relieve his relatives of responsibility.

Alexander then quoted Bernal's explanation of how medicine changed its perspective:

> The beginnings at first were merely a subtle shift in emphasis in the basic attitude of the physicians. It started with the acceptance of the attitude, basic in the euthanasia movement, that there is such a thing as life not worthy to be lived. This attitude in its early stages concerned itself merely with the severely and chronically sick. Gradually the sphere of those to be included in this category was enlarged to encompass the socially unproductive, the ideologically unwanted, the racially unwanted and finally all non-Germans. But it is important to realize that the infinitely small wedged-in lever from which this entire trend of mind received its impetus was the attitude toward the nonrehabilitable sick.
>
> It is, therefore, this subtle shift in attitude that one must thoroughly investigate.[92]

91. Lifton, *The Nazi Doctors: Medical Killing and the Psychology of Genocide*.

92. Alexander, "Medical Science under Dictatorship," *New England Journal of Medicine* 241 (1949): 39, 41 and 41n.17.

The key point to be made here is that the rhetorical and legal foundation has already been laid for a recognition of a moral—if not a legal—duty to encourage death. Daniel Callahan has recently written that "as a culture we need a supportive social context for aging and death, one that cherishes and respects the elderly while at the same time recognizing that their primary orientation should be to the young and the generations to come, not to the welfare of their own group."[93] Should the elderly not readily agree that their orientation "should be to the young and the generations to come," Callahan proposes limiting the legal protection and medical services that will be available to those who have exceeded what he calls "a normal 'life-span opportunity range'" of seventy-five to eighty years of age.[94]

1. The Right to Die

To pose the issue in terms of a stark "duty to die" is not entirely accurate. Rather, the issue is more properly posed as the question of whether *others* (including government officials) have a duty to assist persons who want to die, or to assist those whose lives are considered not worth saving. The legal question identified in the introduction to this paper was stated as follows:

> Should the prohibitions of the criminal and civil law be relaxed to permit active and passive steps to be taken with the specific intent of ending one's own or another's life when that life has lost meaning, quality, or value to oneself or others? At a more fundamental level, the question is this: At what point during the continuum of biological human existence from conception until death should the law impose a *duty* to protect and preserve an individual's life?

Thus, in the case of competent adults who want to die, the legal issue is whether physicians and medical personnel have a legal duty to provide either the means or the necessary assistance.[95] If so, it will be necessary to relax laws prohibiting homicide

93. Callahan, *Setting Limits: Medical Goals in an Aging Society* (New York: Simon & Schuster, 1987), pp. 26-27.
94. Callahan, *Setting Limits*, pp. 136, 171-72.
95. See, for example, *Bouvia v. Superior Court*, 179 Cal. App. 3d 1127, 225 Cal. Rptr. 297 (1986) (*Bouvia I*); *Bouvia v. Glenchur*, 195 Cal. App. 3d 1075, 241 Cal. Rptr. 239 (1987) (*Bouvia II*); *Ross v. Hilltop Rehabilitation*

and assisted suicide.[96] In the case of those who are incompetent, the legal issue is whether others, charged with the duty to act in the "best interests" of these individuals, will be permitted to consent to their death through either active or passive measures because they are, objectively, "better off dead."[97] If so, guardianship and child-welfare standards governing the duty to provide medical care must be relaxed accordingly.[98]

The language of duty already appears in the cases involving incompetent adults. If the incompetent person has a constitutional or common-law "right" to refuse medical treatment, the law must be modified to permit that choice.[99] For the state to hold otherwise would breach what the courts have held to be the legal duty of both medical personnel and the state to permit "death

---

Hosp., 676 F. Supp. 1528 (D. Colo. 1987); *In re Rodas*, No. 86 PR 139 (Mesa County, Colo.; filed 22 Jan. 1987, amended 3 Apr. 1987); *Matter of Farrell*, 108 N.J. 335, 529 A.2d 404 (1987).

96. See, for example, 17-A Maine Rev. Stats. Ann., Sec. 106(6), 201(1)(C), 204 (1983) (authorizing the use of force to prevent suicide, defining suicide by force, duress, or deception as murder, and prohibiting assisted suicide, respectively).

97. *Bowen v. American Hosp. Ass'n*, 410 U.S. 610 (1986) (disabled infants and the duty of recipients of federal funds under Section 504 of the Rehabilitation Act of 1973); *Rasmussen v. Fleming*, 154 Ariz. 207, 741 P.2d 674, 689-91 (1987) (duty of the court and guardians in cases involving incompetent adults).

98. See, for example, *Bowen v. American Hosp. Ass'n*, 410 U.S. 610 (1986) (Section 504 requires no overseeing of non-treatment decisions involving handicapped persons); *Rasmussen v. Fleming*, 154 Ariz. 207, 741 P.2d 674 (1987) (no oversight hearing unless there is a dispute); *Matter of Farrell*, 108 N.J. 335, 529 A.2d 404 (1987); *Matter of Peter*, 108 N.J. 365, 529 A.2d 410 (1987). But see also *Brophy v. New England Sinai Hospital, Inc.*, 398 Mass. 417, 497 N.E.2d 626 (1986) (requiring a hearing); *Superintendent of Belchertown State School v. Saikewicz*, 373 Mass. 728, 758-59, 370 N.E.2d 417, 434-35 (1977) (same).

99. At this writing, the courts have yet to decide the exact basis of the "right to die." See, for example, *Rasmussen v. Fleming*, 154 Ariz. 207, 741 P.2d 674 (1987) (a constitutional right); *Bouvia v. Superior Court*, 179 Cal. App. 3d 1127, 225 Cal. Rptr. 297 (1986) (Bouvia I) (hinting at a constitutional basis); *Bouvia v. Glenchur*, 195 Cal. App. 3d 1075, 241 Cal. Rptr. 239 (1987) (Bouvia II) (indicating a common-law basis); *In re Guardianship of Grant*, 109 Wash.2d 545, 747 P.2d 445 (1987) (indicating a common-law basis).

with dignity," notwithstanding legislative views[100] or individual views to the contrary.[101]

2. The Parental Right to a "Normal" Child

That the existence of a right to end a life which lacks the quality deemed to be a prerequisite to its protection by the law can easily become a duty to assist in ending that life is also illustrated by the rapid progression from the right of the physician's role in abortion to the physician's dutiful involvement as a necessary precondition for effectuation of individual choice. Known as "wrongful birth" cases, these holdings impose a duty on the physician to warn prospective parents about, and hence to test for, potential genetic defects in their unborn children.[102]

The theory in wrongful-birth actions is that the essence of the abortion right recognized by *Roe v. Wade*[103] is the right to make

100. Several recent cases indicate that the courts will not be restrained by clear legislative decisions to proceed cautiously in this area. In both *In re Guardianship of Grant*, 109 Wash.2d 545, 747 P.2d 445 (1987) and *In re Gardner*, 534 A.2d 947 (Me. 1987), it was recognized that the state legislature had explicitly rejected denial of food and water (nutrition and hydration) as an option in "advance directives" cases (e.g., living wills). The courts simply ignored the legislature's concern and made it quite clear that they would go to great lengths to substitute their judgment on the matter for that of the legislature. Compare *In re Estate of Prange*, 166 Ill. App. 3d 1091, 520 N.E.2d 946 (1988), vacated without opinion, 121 Ill. 2d 570, 1988 WL 5111 (19 May 1988) (slip copy) (noting that the Illinois Living Will Act provides that it does not "impair or supersede any legal right or legal responsibility which any person may have to effect the withholding or withdrawal of death delaying procedures in any lawful manner," quoting Ill. Rev. Stat., 1985, ch. 110 1/2, para. 709(d), as amended by Public Act 85-860, effective 1 Jan. 1988 (1987 West Ill. Legis. Serv. 995).

101. See, for example, *Matter of Jobes*, 108 N.J. 394, 529 A.2d 434 (1987) (requiring a nursing home to participate in the starvation death of Nancy Ellen Jobes, notwithstanding the objections of both the home and its staff). But see *In re Guardianship of Grant*, 109 Wash.2d 545, 747 P.2d at 456n.6.

102. See, for example, *Robak v. United States*, 658 F.2d 471 (7th Cir. 1981) (Alabama law); *Turpin v. Sortini*, 31 Cal.3d 220, 182 Cal. Rptr. 337, 643 P.2d 954 (1982); *Fassoulas v. Ramey*, 450 So.2d 822 (Fla. 1984); *Blake v. Cruz*, 108 Idaho 253, 698 P.2d 315 (1984); *Berman v. Allan*, 80 N.J. 421, 404 A.2d 8 (1979); *Speck v. Finegold*, 497 Pa. 76, 439 A.2d 110 (1981); *Dumer v. St. Michael's Hosp.*, 69 Wis.2d 766, 233 N.W.2d 372 (1975).

103. *Roe v. Wade*, 410 U.S. 113 (1973).

an informed choice about whether to end a pregnancy. Thus a physician must disclose the potential disabilities that might affect the unborn child in sufficient detail so as to provide enough information to make the choice of whether to end its life "informed." When states have mandated the disclosure of information about the physical characteristics of normal fetuses in the hope that knowledge of the physical characteristics of the unborn might influence the decision of whether to abort at all, the courts have ruled uniformly that such information *interferes* with the right to choose abortion because it is designed to prejudice the choice in favor of childbirth (i.e., life).[104] From this, it follows that the courts have concluded that prenatal physiological characteristics are not relevant to the woman's choice to have the child unless there is a strong possibility that the child may be disabled. Then and only then do the courts permit the state to require that detailed physiological information about fetal characteristics be disclosed.

In *Blake v. Cruz*,[105] for example, the Idaho Supreme Court noted that although several courts have rejected claims against doctors based on the birth of normal, healthy children, "no court has rejected claims involving children with birth defects" because "society has a vested interest in reducing and preventing birth defects."[106] The woman's right to choose to abort has thus been transformed into a judicially enforceable civil duty *of physicians* to "prevent birth defects" by counseling the abortion of children with handicaps.[107]

---

104. See, for example, *Thornburgh v. American College of Obstetricians and Gynecologists*, 476 U.S. 747 (1986); *City of Akron v. Akron Center for Reproductive Health*, 462 U.S. 416, 442-49 (1983).

105. *Blake v. Cruz*, 108 Idaho 253, 698 P.2d 315 (1984).

106. See *Blake v. Cruz*, 108 Idaho at 244, 698 P.2d at 318-22 (1984); in accordance with *Fassoulas v. Ramey*, 450 So.2d 822 (Fla. 1984) (the cost of raising a normal child is unavailable as damages, but the excess cost of raising a disabled child is recoverable).

107. However, most courts have balked at accepting the theory of "wrongful life" which would impose a duty upon a physician, and perhaps upon the parents as well, to assure that a handicapped child is not even *born*. Under this theory, the quality-of-life reasoning is explicit. See notes 60 and 61.

CONCLUSION

That the courts have largely avoided direct and meaningful discussion of the implications of the "value of life" ethic which drives their holdings bears stark witness to their difficulty. Although the courts simply are not equipped to decide such grave issues of public policy, they do so because the cases are presented to them. When they do, the inevitable result of common-law judicial preference for "case by case" analysis will have a significant effect on the direction of both statutory and constitutional law over the long term.

It was once possible for me to write that "the core ethical issues—those which relate to the 'value' of the life at stake—have been resolved by a tacit agreement to disagree on the ultimate question: whether the law has the right to base the protection it affords upon a subjective evaluation of the 'quality' of an individual's life." But I am no longer so certain of this. Recent case law appears to demonstrate that judicial approval of value-of-life judgments has become the norm, and will shortly become overt. True, the focus in the "hard cases" reaching the courts generally remains on who shall make the decision rather than what decisions they should make, and on the most effective method by which to resolve the difficult medico-legal problems involved. Nevertheless, there can be little doubt that the courts are making precisely the core value-of-life judgments they so earnestly claim to be avoiding.[108]

Unfortunately, such attempts to avoid the inevitable do not work in the life of the law, for the march of *stare decisis* is inexorable. Today's implicit and explicit assumptions drive tomorrow's decisions on the same or similar topics, and the law rapidly begins dealing with issues which only a few years before would have been considered unthinkable as candidates for judicial resolution. Justice Andersen of the Washington Supreme Court has written,

> As recently as five years ago, or perhaps three, the idea that fluids and nutriment might be withdrawn, with moral and perhaps legal impunity, from dying patients, was a notion

108. See *Matter of Peter*, 108 N.J. 365, 529 A.2d at 433 (O'Hern, J., dissenting).

that would have been repudiated, if not condemned, by most health professionals. They would have regarded such an idea as morally and psychologically objectionable, legally problematic, and medically wrong. The notion would have gone "against the stream" of medical standards of care. [However,] . . . this practice is receiving increased support from both physicians and bioethicists. This new stream of emerging opinion is typically couched in the language of caution and compassion. But the underlying analysis, once laid bare, suggests what is truly at stake: That for an increasing number of patients, the benefits of continued life are perceived as insufficient to justify the burden and cost of care; that death is the desired outcome, and—critically— that the role of the physician is to participate in bringing this about.[109]

The steps are incremental, but they go only in one direction. In less than twenty years, the law has progressed from sanctioning the occasional eugenic abortion to abortion on demand, from the disconnection of life-support systems in Karen Quinlan's case,[110] to the reported request in the Hector Rodas case for authorization of a lethal injection.

As predicted, it was merely a matter of time before death by injection came to be regarded as the more humane alternative to death by starvation and dehydration. Although Hector Rodas died before the Colorado court could consider the question raised,[111] it has been reported privately (the court records were sealed) that Rodas's ACLU attorneys requested a lethal injection. The *Journal of the American Medical Association* caused a major controversy over the legitimacy of active euthanasia by lethal injection when it printed an unsigned article from a physician entitled "It's Over, Debbie," in which the physician admitted that a young woman with cancer had been killed with an overdose of morphine.[112]

109. *In re Guardianship of Grant,* 109 Wash.2d 545, 747 P.2d 445, 459 (1987) (Andersen, J., concurring in part and dissenting in part).

110. *In re Karen Quinlan,* 70 N.J. 10, 355 A.2d 647 (1976).

111. *Ross v. Hilltop Rehabilitation Hosp.,* 676 F. Supp. 1528 (D. Colo. 1987); *In re Rodas,* No. 86 PR 139 (Mesa County, Colo.; filed 22 Jan. 1987, amended 3 Apr. 1987).

112. "It's Over, Debbie," *Journal of the American Medical Association* 259 (8 Jan. 1988): 272. See also V. Cohn, "Saving Lives, Ending Lives: Doc-

Doctors in the Netherlands are actively involved in direct euthanasia, protected by court decrees which allow it when a patient makes an informed request.[113] A recent survey of 2,218 Colorado physicians concluded that 60 percent of all doctors have attended patients for whom they believe active euthanasia would have been justifiable if legal, and 58.9 percent of these physicians "indicated that they would have personally been willing to administer a lethal drug if such measures were allowed by law."[114]

In 1988 California voters were asked (though they refused by a wide margin) to approve a ballot initiative that would legalize "death assistance" (including lethal injections under certain circumstances).[115] Similar initiatives are planned for Florida,

---

tors Confront a Mercy Killing," *Washington Post*, 1 Mar. 1988, Health Section, pp. 1, 13 (reprinting *JAMA* article). But see also Letters and Commentaries, *Journal of the American Medical Association* 259 (8 Apr. 1988): 2094-98, 2139-42.

113. See G. E. Pence, "Do Not Go Slowly into that Dark Night: Mercy Killing in Holland." (Pence estimates that, as of late 1987, between 5,000 and 8,000 patients have been killed by physicians.) A summary of the situation in the Netherlands, France, Denmark, the Federal Republic of Germany, and Sweden is contained in a report of the British Medical Association entitled *Euthanasia: Report of the Working Party to Review the British Medical Association's Guidance on Euthanasia*, pp. 49-52. With respect to the Dutch, paragraph 210 of the report concludes that "it therefore seems that, although certain members of Dutch society are against active termination of life for cogent reasons, there is a widespread use of active termination of life, which is motivated by the highest humanitarian ideals but not all of which is reported." The BMA's position on the topic of euthanasia was unequivocal: "The active intervention by anybody to terminate another person's life should remain illegal. Neither doctors nor any other occupational group should be placed in a category which lessens their responsibility for their actions" (conclusion 4, p. 67).

114. Center for Health Ethics and Policy, Graduate School of Public Affairs, University of Colorado at Denver, *Withholding and Withdrawing Life-Sustaining Treatment: A Survey of Opinions and Experiences of Colorado Physicians* (May 1988), p. 16. The Center recommended that the governor should appoint a commission to evaluate, among other things, whether the substantial number of physicians holding these views represent "changing community mores" and urged re-evaluation of present Colorado law forbidding active euthanasia to determine whether it is consistent with "the perceptions of right and wrong" of the people of the state (p. 21).

115. "Euthanasia Law Fails to Qualify for Ballot," NEXIS PR Newswire, 10 May 1988 (Hemlock Society news release).

California (again), Oregon, and Washington in 1990,[116] and one Florida court was recently asked (but refused) to rule that intent to commit euthanasia was a defense to a charge of premeditated murder. The offense: the convicted murderer had fired two bullets into the brain of his wife, who had the misfortune of being afflicted with osteoporosis and the early stages of Alzheimer's disease.[117]

It therefore is well past the time for those charged with the moral guidance of society to step back, to observe the situation, and to provide a clear and unequivocal message about the morality of what has been—or will be—proposed in pursuit of "the good life." It will be insufficient for today's moral and religious leaders merely to seek exemptions for their adherents. That was tried before, and it failed.[118] A more direct approach is clearly needed.[119] We are on the verge of a "Brave New World," and the citizenry, as well as its lawyers and judges, will need all the moral guidance they can get.

116. UPI, "Right-to-Die Group Targets Florida in Ballot Drive," NEXIS, 1 Aug. 1988 (noting Hemlock Society plans to seek assisted suicide initiative in Florida, California, Oregon, and Washington).

117. *Gilbert v. State*, 487 So.2d 1185, 1190-91 (Fla. App., 4th Dist., 1986). Roswell Gilbert's case was recently the subject of a made-for-TV movie.

118. See Lifton, *The Nazi Doctors: Medical Killing and the Psychology of Genocide*. (Lifton notes that at the initial stages of sterilization of the German eugenics campaign, the Catholic Church contented itself with obtaining religious exemptions from participation.)

119. *Withholding and Withdrawing Life-Sustaining Treatment*, the report by the Center for Health Ethics and Policy of the University of Colorado at Denver, notes that "responses from Catholic doctors, more than Protestant or Jewish physicians, suggested hesitation to justify or engage in active euthanasia. Sixty-one (60.9) percent of Protestant, 66.8 percent of Jewish, and 44.0 percent of Catholic physicians have cared for patients for whom they believed legal active euthanasia to be justifiable. Fifty-eight (57.6) percent of Protestant, 61.8 percent of Jewish, and 47.5 percent of Catholic doctors acknowledged that they would administer a lethal dose to such patients if it were legal."

God, the Body, and the Good Life

Paul C. Vitz

Anxiety in the face of an autonomous, ethically out-of-control science and technology is probably the most pervasive theme of our conference. The philosopher Thomas Molnar, after reflecting on various recent responses to this problem, put it this way: "Mankind faces technology as it once faced nature itself, that is, as an overwhelming power against which protection must be sought."[1]

The problem, then, is how to protect ourselves against this new power. In *The Imperative of Responsibility* Hans Jonas proposes that we find lessons "in archaic conceptions of the world order and man's place within it." What we need, he writes, is an "heuristic of fear" and an "ethic of cosmic responsibility" based on humans taking responsibility for nature's integrity in a manner analogous to our taking responsibility for human cultural institutions.[2]

But Molnar thinks that this call for a "salutary fear and an ethics of responsibility" will inevitably fail, since today humans know that there are no internal constraints upon the use of technical power; after all, for over two hundred years the tendency

1. Molnar, "Technology and the Ethical Imperative," *Chronicles* 12 (Feb. 1988): 14.
2. Jonas, *The Imperative of Responsibility: In Search of an Ethics for the Technological Age* (Chicago: University of Chicago Press, 1984), cited by Molnar in "Technology and the Ethical Imperative," p. 14.

has been to remove constraints in favor of efficiency. Molnar goes on to say that "all that our age is able to produce is an ethics of technology which is not a moral reflection *about* technology but a method of finding ways to adjust to it."[3]

Behind it all Molnar points to the mechanization of man, a mechanization that is itself a product of science. Thus Molnar concludes that "the truth is, one cannot harness technology without demoting science as a master-concept." Yet Molnar despairs of such a demotion, for he claims that it would require change so radical that it is literally inconceivable within Western discourse, since our basic vocabulary and concepts are shaped by meanings which at present are all geometrical and mechanical.

Perhaps such a conclusion is right, and a somber future of technological madness is slouching inevitably in our direction. Certainly Western intellectuals find it much easier to describe and diagnose the malaise than to chart any direction for escape. The risk of suggesting an answer to the dilemma is that of being labeled moralistic or ridiculous or naïve—or all three. Nevertheless, I will offer such a proposal here, but first I will sketch a somewhat different perspective on the problem.

Along with many others, I see the essential problem not so much as science itself but as the reductionistic, individualistic philosophy and worldview that lie behind today's science. In particular I will focus on the well-known assumption that the self has no essential relationship either with the body or with transcendent reality—specifically, God. That is, the modern self or mind has rejected any serious connection to two of the three major realms previously understood to exist outside of it. (The third major realm is that of other persons; a good case can be made that the modern self also rejects this reality, but that will not be of direct concern here.) Since this existential self is morally relative and since it exists suspended in a kind of perpetual subjectivity, nothing has any convincing degree of meaning—a meaning capable of providing significant moral restraint. In its quest for autonomy from nature and from God, the modern self has created a world in which one man's narcissism is another man's nightmare. The answer to this dilemma, let me suggest, is to restore the connections between the modern self's alienated mind and its body, and

---

3. Molnar, "Technology and the Ethical Imperative," pp. 14, 15.

between this same mind and God.[4] I will make a particular case for both kinds of restoration. Although such a proposal may be foolhardy, it clearly fits Jonas's previously noted proposal—that we find lessons in ancient conceptions of the world and of man's place in it. Nothing so characterizes "archaic" world orders as the belief that the body is part of the self—in fact, the body is often sacred—and the belief that man's meaning is closely tied to God, or the gods.

I will start with the modern mind or self and its separation from the body, and I will treat this well-known split within the context of recent discussions of Artificial Intelligence (AI). I believe that theory and research in Artificial Intelligence both fully expose the nature of this modern mind/body split and show that this split is scientifically—that is, empirically—false. Here I agree with the view that science and technology can illuminate our moral dilemmas—an illumination that comes about when science and technology provide a greater understanding of what human nature actually is.

THE MIND/BODY SPLIT

Angels, whether one believes in them or not, are defined as rational beings without bodies. By contrast, we humans are known to be rational beings *with* bodies. In a standard Judeo-Christian framework, humans are not only rational beings who happen to have bodies; they are embodied rational beings. That is, human mental life and human bodily life are theologically conceptualized as inextricably interwoven. On the other hand, the Greeks and their modern idealistic heirs see the human mind as only accidentally connected to the body. For those taking this approach, we humans are rational beings who just happen to have bodies— but our bodies are only a necessary accident of our having a physical existence and are not intrinsic to who and what we are.

4. This is hardly a novel analysis. For example, in 1835 Catherine Beecher described the modern dilemma: "The great crisis is hastening on, when it shall be decided whether disenthralled intellect and liberty shall voluntarily submit to the laws of virtue and of Heaven, or run wild to insubordination, anarchy, and crime." I interpret virtue and character to be a kind of morality heavily based on the body and on everyday experience.

For these theorists, there is no necessary link between the nature of our body and the nature of our mind.

I still remember when I was first introduced to the concept of a computer program. It was about 1960, and I was a graduate student in psychology. My professors emphasized that the beauty and power of a program lay in its independence from the particular physical material in which it might exist. A program, like a statement in formal logic, can take various forms: it can be written in chalk on a blackboard, it can exist as a sequence of ones and zeros as written in machine language, it can be punched as holes in a deck of IBM cards, it can be a magnetic pattern on tape, or it can be a sequence of electronic states in the computer itself such that the program can be stored and then retrieved and run. In other words, a program can be embodied in almost any material so long as the material in question allows one to fix the symbols expressing the program; the program with its structure exists independent of any particular physical medium. Strange as it may sound, a computer program is somewhat closer to an angel—that is, to a rational intelligence without a body—than it is to the mind of a human being—at least that is the claim being made here.

This understanding of the mind, which is found in so much of the Artificial Intelligence world, has a very clear philosophical pedigree. The philosopher H. L. Dreyfus notes that the mind/body separation begins with Descartes and continues through Hobbes, Kant, and Leibniz.[5] More recent representatives of this tradition so strongly influencing AI include Gottlob Frege, Edmund Husserl, Bertrand Russell, and the early Wittgenstein. Dreyfus points out that the classic AI research program of Alan Newell and Herbert Simon and that of Marvin Minsky represent empirical tests of this disembodied conception of mind. That is, this well-known rationalistic and idealistic philosophy of mind has become the central research program of AI over the last thirty years, and thus has been subjected to a vigorous empirical test. Furthermore, this theory is now understood to have failed in two

5. See H. L. Dreyfus, "From Micro-Worlds to Knowledge Representation: AI at an Impasse," in *Mind Design*, ed. J. Haugeland (Cambridge: MIT Press, 1981), pp. 161-204; and H. L. Dreyfus and S. E. Dreyfus, "Making a Mind versus Modeling the Brain: Artificial Intelligence Back at a Branchpoint," *Daedalus* 117 (Winter 1988): 15-43.

major ways. First, this understanding of mind has failed the biological test of the nature of mind, as I will subsequently explain. But it has also failed within the world of AI itself.

What Dreyfus documents clearly is that the AI researchers themselves have discovered that their basic assumptions don't work. They don't work because it is not possible to construct a formal theory—a set of rational statements—to represent large amounts of human knowledge of a practical or bodily type. Dreyfus calls this "commonsense knowledge." There is no formal theory for commonsense physics—for example, moving a piano. In addition, Dreyfus makes it clear that both Heidegger and Wittgenstein (in his late work) understood the dilemma of commonsense knowledge and identified it clearly. (Dreyfus also suggests that our commonsense knowledge of people is incapable of formal representation. This very difficult task hasn't even been given much attention by AI people. Interpersonal knowledge, which is so central to what is meant by the human mind, is probably still "more" impossible to represent than commonsense physics. After all, the physical world doesn't have "a mind of its own.")

What, very briefly, is the nature of the failure? Dreyfus describes it from two perspectives.

First, it turns out that a person's intuitive commonsense knowledge is incredibly hard to formalize, even for very simple tasks. Personal comments from various experts have been telling. For example, a child pushing a Playdoh-like substance into a cardboard tube with a hole in the bottom soon learns that the Playdoh will bulge out of the hole; the child soon learns that sand, water, and rocks will "behave" differently. Attempting to formalize this rather simple kind of practical physics took a small team of experts all summer—and even then they failed! It turns out that there is no good evidence that practical physical knowledge has a theory; instead, it is simply learned, but it can't be formalized—not without contradictions, exceptions, and all kinds of qualifications that the programmer usually will not think of in advance. This same problem is considered even more enormous when it comes to trying to represent the linguistic knowledge of even a four-year-old child.

From another perspective it turns out that a person's purposes are always changing the meaning of the properties of the physical world. As a result, AI researchers have discovered that there are no fixed abstract elements or fixed relationships among them. At least not for everyday, commonsense, bodily knowledge as distinct from scientific knowledge.

In short, evidence now strongly suggests that the classical, symbol-based AI research program representing the rationalist and reductionist tradition has failed. "The idea of producing a formal, atomistic theory of the everyday commonsense world and of representing that theory in a symbol manipulator" is now an exhausted, dwindling research program.[6]

In addition to failing on the preceding grounds, the separation of mind and body also fails on biological grounds. In fact, the fundamental difference between a computer program and the human mind has long been established empirically in the biological sciences. And in the past few decades, research in neurophysiology has very thoroughly elaborated and deepened the evidence that the human mind is dependent on the different particular materials of the brain. The research is well known, though apparently the implication—that computer programs are quite different from the human mind—is not commonly appreciated. This biological evidence constitutes the second major demonstration of the failure of the rationalist model of the mind.

Over 150 years ago, the great German physiologist Johannes Müller first clearly articulated what is known as the "law of specific nerve energies." Put simply, this means that a given nerve gives rise to a sense quality that depends on the specific character of the nerve—that is, the stimulation of a visual nerve gives rise to visual experience, the stimulation of an auditory nerve gives rise to the experience of sound, and so on. We can hear, for example, because there are specific nerve fibers in the cochlea for almost each specific sound frequency. Thus the hair cells on the organ of Corti at the bottom of the cochlea respond to high frequencies, while those at the top respond to low frequencies. Now Müller's principle is far more general than the qualitative experience of the five senses, because this principle characterizes the central nervous system—that is, the cortex as well as the peripheral senses. For example, recent research shows that this kind of qualitative specificity is present in the auditory cortex,

---

6. Dreyfus, "From Micro-Worlds to Knowledge Representation: AI at an Impasse," p. 34.

where it is known as a tonotopic map.[7] That is, the frequencies to which the hair cells in the cochlea are sensitive are mapped into columns of cortical cells—with each column of cells responding only to a particular and very narrow band of tone frequencies. In short, the particular neurons in the auditory cortex are not interchangeable, general-purpose neurons like silicon chips; rather, they are highly specific and qualitatively different.

This same specificity principle characterizes the visual system—indeed, with sight the degree of specificity is, if anything, even greater than that of audition.[8] It has long been known that in the retina there are three different kinds of color-sensitive receptors (cones) plus light-sensitive receptors (rods). However, research that began three decades ago has demonstrated that retinal ganglion cells (the best-identified types are known as X, Y, and Z cells) are also specialized for certain elementary kinds of light stimulation as well as for location on the retina; in the lateral geniculate nucleus (part of the brain), visual neurons are specialized for one of four colors, for location on the retina, and so on. In the visual cortex there is even greater specialized complexity. Here we find groups of visual neurons specialized for straight lines of different orientations ranging from vertical to horizontal. Further, it appears that there are cortical visual neurons that respond only to binocular disparity, other groups of cortical neurons that deal only with color processing, and still other systems that specialize in the perception of form and movement. In short, throughout the structure of the visual cortex there are qualitatively distinct channels analyzing and responding to elementary visual properties. Typically these channels process the various qualitatively different kinds of visual information in parallel—that is, simultaneously.

Since the mid-nineteenth century it has been known that elsewhere in the cortex there are special systems both for understanding speech (Wernicke's area) and for producing speech

7. See G. L. Romani, S. J. Williamson, and L. Kaufman, "Tonotopic Organization of the Human Auditory Cortex," *Science* 216 (1982): 1339-40.
8. The information that follows can be found in most good recent treatments of the nervous system—for example, E. R. Kandel and J. H. Schwartz, eds., *Principles of Neural Science*, 2d ed. (New York: Elsevier, 1985).

(Broca's area). The motor cortex is another major area of specialized neurons. Indeed, the cortex is now known to consist of a very
large number of interconnected subsystems of neurons, each with
specialized, qualitatively different sensitivities. One major consequence of this now-established understanding of the cortex is that
to simulate the human mind it will be necessary to simulate the
human nervous system.

This extension of the law of specific nerve energies from the
peripheral sensory system to the cortex clearly shows that the
human brain operates on a principle which is the opposite of that
of a digital computer. That is, digital computers are made of identical and interchangeable electronic elements. The possibility that
certain chips could process only one kind of information—for example, a payroll but not a mathematical equation or a business
letter—would destroy the utility, the very *raison d'être* of the
modern digital computer.

This is not to imply that all cortical neurons are qualitatively different from each other. Certainly within a cortical neural system there is some redundancy; for example, a whole column of
cells in the visual cortex may be sensitive to the same line orientation (or spatial-frequency orientation). But such local redundancy should not be allowed to keep us from understanding that
many different cortical areas are involved in qualitatively different kinds of processing and experience.

In other words, the current understanding of the cortex is
that it consists of a complex, interconnected group of subsystems.
Each of the many subsystems represents a specialized and
qualitatively different kind of processing; often these subsystems
are also associated with qualitatively different *conscious* experience. All this means that the basic neural elements—or the
"chips"—in each subsystem would have to be highly specific and
different from those in every other subsystem; the same is also
probably true for the large number of interconnecting neural
structures. There is what might be called "a law of specific neuron
systems."

A different but closely related fundamental biological fact is
that the nervous system and the human body as well are intimate-
ly linked with properties of the external physical world. Consider
just one example: the range of light waves to which the human
eye is sensitive. This range, known as the visible spectrum, is from

about 380 nm (violet) to 760 nm (red). Now the potential spectrum of light (electromagnetic energy) is enormously greater and ranges from extremely short waves—for example, gamma rays—to very long radio waves and AC circuits. The visible spectrum is thus a very small slice of this potential spectrum. However, it is reasonable to assume that the human eye is concerned only with the light available on the surface of the earth. To be able to see waves that exist only elsewhere in the cosmos would be a waste of biological energy, tissue, and so on. But the human eye is responsive to almost all of the spectrum that actually reaches the surface of the earth with any significant amount of energy; only the relatively small ultra-violet and infra-red parts of the spectrum are not part of the human eye's sensitivity. That is, the human visible spectrum is close to the available spectrum on the surface of the earth.

Over and over again scientists find evidence of this kind showing how the body is adapted—even fine-tuned—to its environment. The origins of this intimate connection are well understood within the context of evolution, whether evolution is viewed atheistically and derived from chance or is viewed theistically and derived from God. That is, both approaches assume that life in all its forms is closely connected to the external environment in which living things have developed and to which they are adapted. Both kinds of scientists assume that an animal's nervous system can't be understood when separated from its body and that neither the nervous system nor the body can be understood when separated from the animal's environment, since the three constitute a mutually interacting system.[9]

The major point that the mind is embodied is, of course, hardly a new one—but it is a point still being made. Recently it has been emphasized in the writings of the information theorist Donald MacKay[10] and in the critiques of artificial intelligence by

9. For a recent treatment of the remarkable way in which the properties of the physical universe are coordinated with the requirements of life, see G. Greenstein, *The Symbiotic Universe* (New York: Morrow, 1988).

10. See MacKay, "Mind Talk and Brain Talk," in *Handbook of Cognitive Neuroscience*, ed. M. Gazzaniga (New York: Plenum, 1984), pp. 293-317; and "Machines, Brains, and Persons," *Zygon* 20 (1985): 401-12.

Dreyfus, who sums up his position with a quote from the poet Yeats: "Man can embody the truth, but he cannot know it."[11]

I am aware that some of the difficulties that arise from ignoring the body are beginning to receive serious attention in AI and related areas. In other words, to some extent AI has accepted the critiques just mentioned. Neural nets, now fairly common, are a small step toward a more neurological or "body-like" model of the mind. That is, AI has *recently* moved substantially away from the abstract digital approach to simulating the mind and toward recognizing the importance of the body. Nevertheless, very serious difficulties remain to be dealt with before even a modest simulation of the biological basis of the mind appears possible.[12]

One expression of the difficulties involved in the simulation of the human brain is represented by the terms "hardware" and "software." "Hardware" refers to the fixed physical and electronic components in a computer or robot. However, there is no real hardware analogy in the human body, in which even muscle and bone tissue is, at best, a kind of "software." A computer program is called software, but there is no evidence that the program level actually exists for humans. The body exists and conscious experience exists, but there is no evidence that a level analogous to a computer program exists as a functioning part of the brain/mind.

Much of the basic simulation difficulty arises from the fact that computers and robots are based on silicon while animal life is primarily carbon-based. Computers are not in principle restricted to silicon systems, but they are all (including neural nets) based on silicon for what appear to be practical reasons— namely, silicon is cheap and allows very reliable binary operations. As such, silicon systems are devoted to dryness, so to speak, while carbon systems are devoted to wetness. Water quickly destroys or "kills" a computer, while too much dryness quickly kills humans and other animals. The brain is very much a wet system, and simulating it will have to involve simulating this very fundamental property, which is so different from that of com-

11. Dreyfus, "From Micro-Worlds to Knowledge Representation: AI at an Impasse," p. 204.

12. See, for example, J. T. Schwartz, "The New Connectionism: Developing Relationships between Neuroscience and Artificial Intelligence," *Daedalus* 117 (Winter 1988): 123-41.

puters. In brief, the human brain consists of different kinds of what might be called "wetware," and "hardware" and "software" are irrelevant or misleading terms. In any case, my fundamental point here is that a true simulation of the human mind would require a simulation of the human brain and body. Whether this is possible remains to be seen.

But let us return to the original concern: the loss of common values in contemporary life capable of restraining aspects of technology, especially with respect to human biology. To recover an understanding that humans are intrinsically related to the body should be helpful. For example, this principle means that to the extent that new biological properties would be made part of humans, to that extent we would be creating a new type of mind and thus a new type of being. One could ask whether these new so-called humans would in fact be a different species—especially if the method of reproduction were different. This may be too extreme a response, but we should bear in mind that the relatively small biological differences between the presently existing races and various ethnic groups already cause more than enough conflict. New biological engineering that would pit "naturals" against supposedly superior "artificials" might make the now-familiar problem of racism look positively trivial. (One is reminded of the movie *Invasion of the Body Snatchers*.) In short, to make new people who would be biologically different in any significant way would be to create a significantly new type of person and with these differences to create serious possibilities for social conflict.

In any case, the understanding that we are embodied means that the philosophical tradition which assumes that the mind is a basically abstract and disembodied phenomenon has been shown to be wrong—we are inseparable from our bodies. (Jerusalem has been vindicated, if you will.) Still, the acceptance of the embodied nature of the human self or mind is not really powerful enough to overcome the contemporary moral vacuum. It is simply not clear how such an understanding of the body as an inextricable aspect of the mind is an adequate basis for a moral consensus about how to guide science and technology. It may give us more humility, more anxiety about creating new "races" or "breeds" of human beings, and it may be a step in the right direction, but it certainly seems far from enough.

THE MIND/SPIRIT PROBLEM OR
THE GOD QUESTION

Therefore, I turn to the other major issue: the question of God's existence—or, more generally, the question of the reality of the transcendent spiritual realm. With the preceding mind/body material as background and context, it is time to focus on our central topic: the modern separation of the mind and hence of the self from spiritual life.

First, I wish to emphasize that the prior point on the inter-relationship of the mind and the body is proposed as an analogy to a similar interrelationship between the mind and the spirit. Just as the human mind is inextricably bound up with the body and physical reality, so it is likewise bound up with God and spiritual reality. Thus I start with the assumption that there *is* a transcendent spiritual realm, and that the human mind/body is constantly interacting with this realm. Now I am fully aware of the fact that it is precisely this assumption that is rejected by many Western intellectuals and scientists, and not just those in the world of Artificial Intelligence. What I will do, then, is ex-amine the basis of this rejection and present a case for the exis-tence of spiritual reality. For it is obvious that one must first ac-cept this realm as existing before one can accept its relevance for an understanding of the present dilemma. Accordingly, the sub-sequent remarks are primarily addressed to the skeptic or athe-ist. Let me be clear: I am convinced that the God question which loomed so large at the start of modernity when atheism was first clearly formulated must also be dealt with today if the pathol-ogies of modernism are to be overcome. Let me also say that I see the Western intellectual community as showing serious bias (or even worse) in failing to bring the God question into the open and in failing to make it part of the public debate. In some small way I hope that this conference will break this taboo against the public mention of God. For, as I see it, Western intellectuals have begun to wake up among the swine and are ready to begin a prodigal journey. If so, the time should be ripe to give the God question and the spiritual life an active place on the Western in-tellectual agenda. With that said as editorial comment, let us take up the argument.

An important point is that throughout human history and

its varied cultures, three great external realms of reality common-ly have been assumed to exist. These are the external physical world, the world of other minds, and the transcendent spiritual world—for example, of God or the gods. An interesting feature that these three presumed realities share is that we cannot prove the existence of *any* of them. Indeed, some years ago the promi-nent philosopher Alvin Plantinga published a very important proof on the subject. Briefly, what Plantinga was able to prove was that the degree of rational and empirical uncertainty about the existence of other minds and the degree of rational and em-pirical uncertainty about the existence of God are exactly the same. That is, the rational grounds for accepting the existence of both of these realms have the same structure and involve the same assumptions—assumptions that Plantinga shows are often question-begging in both cases. For example, one never directly experiences other minds, and one's assumption that they exist is based on an analogy with one's own mental life.

Plantinga's proof is sophisticated and cannot be summa-rized easily, but its general structure is not hard to outline. First Plantinga systematically shows that neither natural theology nor natural "atheology" offers a satisfying solution to the problem of a rational justification of belief in God's existence or in God's non-existence. He then tries another approach to the justification of belief in God by exploring its analogies and connections with a similar issue, the "problem of other minds"—that is, how do we justify the existence of other people's minds? Plantinga goes on to "defend the analogical argument for other minds against current criticism and argue that it is as good an answer as we have to the question of other minds. But it turns out that the analogical argument finally succumbs to a malady exactly resembling the one afflicting the teleological argument [for God's existence]." Plantinga thus concludes that "belief in other minds and belief in God are in the same epistemological boat; hence if either is rational, so is the other. But obviously the former is rational; so, therefore, is the latter."[13] His formal proof for this conclusion has stood without a successful challenge for over twenty years.

13. Plantinga, *God and Other Minds* (Ithaca, N.Y.: Cornell Univer-sity Press, 1967), pp. vii-viii.

Elsewhere Plantinga shows that just as it is impossible to prove the existence of other minds, so it is also impossible to prove the existence of external physical reality, or even to prove the existence of the past.[14] Again, he shows that the failure in each proof is identical to the failure in the teleological argument for God's existence.

One obvious implication of Plantinga's work is that if scientists, for example, tend to assume the existence of physical reality and of other minds but to reject the existence of God, they do so on non-rational grounds. Before turning to some of the non-rational reasons behind the rejection of the spiritual realm, it will be useful to discuss how it is that the existence of the external world is commonly accepted. The problem of proving the existence of external reality arises once one accepts the position that our knowledge of external reality is always mediated by the nervous system. All we are directly aware of are our own states of mind. We must—we can only—infer an external reality existing behind and acting as a cause of our sensations, our perception, and so on. The validity of this inference is what cannot be proved. We may accept Plantinga's reasoning in this matter, or we may be convinced on other grounds that proving the existence of the physical world is not possible. There is, of course, a long line of skeptics on this issue in Western philosophy—David Hume, Bishop Berkeley, Thomas Reid. Their writings certainly support Plantinga's conclusion on this issue.

Nevertheless, almost no one has ever doubted physical reality to the point of trying seriously to live by such a position. If a person lived on the basis of such doubt, it is not clear why he or she would eat food, would avoid walking into walls, or would even bother to get dressed. A few idealistic philosophers in the last two hundred years or so seem to be the intellectual representatives of a position that does deny or comes close to denying the physical world.

However, the overwhelming majority of scientists and of average citizens of the world have always accepted the existence of an external physical reality. Scientific theories are, after all,

14. Plantinga, "Reason and Belief," in *Faith and Rationality: Reason and Belief in God,* ed. Alvin Plantinga and Nicholas Wolterstorff (Notre Dame, Ind.: University of Notre Dame Press, 1984).

about something outside of us. The ground for this acceptance seems to be that we are so made that sensory and perceptual experience carries with it the overwhelmingly convincing notion that it is external reality that we experience. Put somewhat differently, our normal interaction with what appears to be physical reality naturally creates a firm conviction of its existence.

Of course, in some rare instance one's perception of external reality may be faulty. There are such things as illusions, hallucinations, and so forth. But to believe that the whole realm of physical reality doesn't exist or that most or even much of human perceptual experience is without an external source would be considered—would *be*—bizarre indeed. Except for certain kinds of philosophers such as the just-mentioned idealists—who are given a kind of philosophers' license to suspend common sense—anyone who failed to believe in the external world would be judged as suffering from a mental pathology.

Likewise, our belief in the existence of other minds comes from our interaction with other people. Sensory contact with a person plus interaction involving language and symbols appear adequate for us to reliably assume the existence of other minds. The tendency to interpret other minds as existing is so strong that often it reaches the point of projecting human minds onto creatures that have very different kinds of minds, or projecting mind onto something that has no mind at all. Children project human minds onto many animals; they even project minds onto trees and inanimate objects, especially at night. Anthropologists commonly note that in so-called primitive cultures certain special objects such as masks, talismans, and so forth sometimes are superstitiously understood as possessing mind and spirit. This tendency, known as anthropomorphism, is one that scientists have traditionally guarded against. However, some of those in AI seem especially susceptible to this error of projecting mind onto objects; for example, one prominent AI scientist attributes beliefs to thermostats.[15] Apparently thermostats have three beliefs: "it is too hot," "it is too cold," and "it is just right." That a thermostat has beliefs seems to me to be a rather crude if updated example of anthropomorphic thinking. (Indeed, apparently even the notion

15. See J. Searle, *Minds, Brains and Science* (Cambridge: Harvard University Press, 1985), p. 30.

that a complex computer program is "thinking" is a case of an-thropomorphic thought.[16])

Although even AI scientists may sometimes see mind in or project mind onto things or places where it doesn't exist, few seriously propose that other minds don't exist. Even if mind is as-sumed to be an expression of matter, few doubt that other people's integrated consciousnesses—that is thoughts, feelings, and purposes—actually exist. Practically speaking, everyone assumes both the existence of other minds and the existence of physical reality.

It is important to note that a crucial factor with respect to initiating and maintaining contact with external physical or men-tal reality is the person's willingness or desire to initiate and main-tain the interaction with the presumed reality. For example, sup-pose you find a man who is on an artificial respirator in a darkened room, and he claims there is no external reality. After some investigation, you discover that he has not walked or used his eyes or ears for some time. His last tape-recorded utterance is a comment to the effect that there is no external physical world. You desire to cure him of this intellectual ailment—one obvious-ly supported by his markedly reduced physical and perceptual activity. A reasonable strategy would be to strengthen his muscles, to get him to open his eyes and unstop his ears, and to talk with him often. In time, you as his guide would ask him to walk and later to come out of his room and, if you will, enter the outside world. Therapy for his pathological intellectual position would thus mean immersing him in the direct interactional experience of the reality that he denies. In this case there is every reason to believe that such a program would convince him of the realist position. But such a procedure would depend upon his willing-ness to cooperate with you, and as for proof of that reality, that would remain, as always, impossible.

Suppose you find someone who not only denies that other minds exist but also lives as though other minds don't exist. (Such a position, of course, seems to be quite rare.) Let us also suppose,

16. See, for example, the important observations of W. C. Hill in "The Mind at AI: Horseless Carriage, Mouse Trap, Card Trick, Clock" (Austin, Tex.: Microelectronics and Computer Technologies Corporation, 1988), unpublished ms.

as would be likely, that the subject's condition is strongly supported by his social isolation. He lives alone and has done so for years; he never speaks to anyone—he appears to have withdrawn completely from interpersonal communication. Accordingly, his lack of belief in other minds is hardly surprising. He remembers interacting with people when he was young, but these experiences he attributes to a childish and immature understanding of things at the time. Again, this man's condition is fundamentally a mental pathology, and correction would involve the slow introduction of interpersonal communication into his life. In time he would discover friends (and enemies)—perhaps even love. Later, after he had experienced years of friendship, if an old friend were to remind him of his former belief that other minds didn't exist, he would probably just look at his friend and laugh. In short, one needs to interact with other minds in order to accept their existence; indeed, in almost all cases such interaction is sufficient to produce such acceptance.

Let me suggest that the situation with respect to belief in the transcendent spiritual realm is similar. First note that most of the people who deny not only the existence of God but also the existence of the entire spiritual realm constitute a relatively small group that seems to have come into being in Western Europe about 250 years ago. Today its members live in a rather peculiar environment. Most of them have been trained in science or other rationalistic and intellectual disciplines. They tend to work in laboratories and universities, which we all know are highly specialized and peculiar places. They tend to socialize mostly with those having similar skeptical outlooks. What they mean by "real thinking" is the mental manipulation of abstract written symbols—often numbers or other digital elements. To such people, a proper belief system or worldview is something constructed by the correct sequencing of these symbols, with occasional checks on whether some kind of observation backs it up. That is, their worldview is something that exists in a digital code, and they seem to assume that digital codes are adequate for representing any kind of question, problem, or knowledge. The very notion of a belief system—based on an oral tradition of knowledge, or on analog information coded in the body and often unavailable to conscious verbal expression, or on a worldview based primarily on direct personal experience—doesn't occur to them.

Also "strange" is the fact that these people never or almost never go to church or synagogue or read religious writings. But most peculiar of all is that they appear never to pray, to meditate, or to engage in other spiritual exercises. That is, they rarely if ever use the well-known procedures for getting and staying in contact with the spiritual realm.

Again, the answer to this pathology is not some vain attempt to prove the existence of God or spiritual reality. As in the other cases just cited, this is impossible anyway. The answer is to try to convince such persons to pray—that is, to talk with God or listen for God's voice, or to engage in other spiritual activities. If such people refuse to interact with the transcendent and are determined to remain in their spiritual isolation, there is little else one can do.

This requirement that one engage in prayer and meditation is a serious one. One must make an effort to eliminate doubt. For example, if someone doubted a certain astronomical claim—such as the existence of moons around Jupiter or the reality of a whole level of physical existence (e.g., subatomic particles)—an honest search for an answer would require a number of things. First, if the person were ignorant of astronomy or physics, he or she would need a guide—a trained scientist—and would have to become at least something of an amateur scientist. It would take considerable time and commitment from the seeker. After all, observations are often ambiguous, and in any case, observations don't reliably interpret themselves.

In almost all religious and spiritual traditions, a knowledgeable person—a guide, if you will—is needed. And prayer and meditation are the primary instruments—the telescopes—for contacting and interacting with spiritual reality. No scientist who refuses to seek religious experience has the intellectual right to say that spiritual reality doesn't exist or that the mind cannot be affected by that reality. A person who has had no religious experience is simply unqualified to comment on the existence—much less the nature—of most spiritual phenomena.

One critic has replied that my argument means that an obstetrician, for example, must experience pregnancy and birth in order to be a good practitioner. Since men have been leading obstetric physicians for years, my argument is supposedly unconvincing. But let's look at this example in some detail. First, the actual experience of pregnancy and birth is close enough to some experiences that everyone has had (e.g., swelling, pressure, pain, etc.) that those who have never been pregnant can understand it. Second, there are well-known reasons why a man cannot have the specific experience of pregnancy. (Natural deficiencies limit experience. So perhaps many scientists are constitutionally opaque to spiritual experience like some people are tone deaf?) Third, I claim that if the doctor has not observed many pregnancies and also accepted these experiences as real, then he is *not* a satisfactory obstetrician. It is only by accepting the reality and validity of the normal pregnancy that he can assume that there is indeed a phenomenon known as pregnancy and a field of study that focuses on it. Once the doctor accepts the experiences as real, then he can judge whether a given woman is having a normal pregnancy, an abnormal pregnancy, or (in rare cases) a false pregnancy.

Similar discriminations are also made about spiritual experience. Judgments about spiritual experience may be less reliable than those made about the normality or abnormality of pregnancy, but that is not a reason for rejecting the reality of all spiritual experience. Judgments about many medical conditions have become relatively reliable only in recent years, and this has occurred only because these judgments have been subject to long-term systematic investigation.

Please note that I am not saying that the person must have a particular interpretation or understanding of his or her religious or spiritual experience; I am saying only that he or she must have had a reasonable amount of such experience to reach informed conclusions about it. Perhaps, after various religious experiences, the person will conclude that such experience was all an illusion or something other than what it first appeared to be. Fine. Scientific observations too can be mistaken, and so can particular spiritual experiences—perhaps all such experience is illusory. However, a scientist without a systematic empirical understanding of a phenomenon would not be in a position to give informed criticism. And a scientist who was ignorant of and refused to get involved with the experimental methodology used to demonstrate that a major phenomenon existed would be considered irrelevant to evaluating the claim. If he actively persisted in rejecting the phenomenon on a priori grounds, his colleagues would rightfully dismiss his claims as unqualified—even should subsequent research prove his position to be right.

I trust the argument is clear. For most people religion is supported by religious or spiritual experience in which they claim a relationship or interaction with a spiritual realm. This may mean interaction with God, or with Jesus, or with a dead person, or even with evil spirits. To evaluate the validity of these extremely important claims of interaction, an investigator must seek contact with spiritual reality. There are various ways the investigator can do this—but first he or she must have the will to actively seek. The desire to seek, of course, is something rooted in psychological factors and has relatively little to do with what is usually called "reason" or "evidence." Given the will to seek, then, the most common instruments or techniques for contact with spiritual reality are prayer and meditation; they are, if you will, the "telescopes" of the religious person. No true scientist should be afraid to seek new knowledge or be afraid to look through any kind of telescope. Another way to put it is to note that in the United States alone some 95 percent of the population say they believe in God. Every year millions have important religious experiences. To ignore evaluating this obviously relevant information is an act of intellectual denial that rather boggles the mind.

I present the preceding case for belief in the transcendent realm not just because of its bearing on the intellectual problem of Artificial Intelligence. AI, of course, is involved in the simulation of intelligence, which often means simulating the human mind. The possible existence of mental interaction with spiritual reality—in particular, with God—relates to this task. Of course, if God and other spiritual "persons" or forces are purely psychological phenomena that are projected into "heaven," so to speak, then no additional or new problem is involved. But if, for example, God exists, and if some of the people are some of the time doing God's will and not their own will, then the problem of simulating human mental life takes on even more serious difficulties than those already noted, to put it mildly.

The major reason for raising the God question, however, is because of its relevance to the moral dilemmas of our technological age. These dilemmas took years to unfold, and no doubt they will take years to resolve. I am convinced that until we humans can regain the perspective of our *intermediate* place in reality—between God and nature, inextricably connected to both—we have no way to deal with our moral crisis. Our problem today stems from a narcissism derived from an intellectual misunderstanding of who we are. We are not autonomous; we are part of our bodies, part of nature, and yet our meaning and purpose are transcendent, derived from God. If we are once again to accept or even to confront this ancient wisdom—a wisdom necessary for anything like a truly "good life"—the Western intellectual community will have to address the God question.

The Story of an Encounter

Richard G. Hutcheson, Jr.

THE "NEW CLASS" AS GUARANTORS OF "THE GOOD LIFE"

The conference entitled "Guaranteeing the Good Life" was reconvened for its second meeting at the Union League Club in New York City in late May 1988. The roster of participants was essentially the same. The earlier meeting had begun with a consideration of the philosophical basis for moral judgments having to do with the good life. The second conference began with a sociological perspective presented by Brigitte Berger of the sociology department of Wellesley College (and now of Boston University).

In introducing the opening session, moderator Richard Neuhaus asked why certain ideas about "the good life" and about what ought to be done to advance or even guarantee it have the degree of cultural sway that they have today. "Again we're talking about everything from abortion, to reproductive technology, to all the other ways in which there are cultural and public-policy disputes about smoothing the rough edges and removing the contingencies of life," he said. "Part of it, as Brigitte makes clear in her argument, is the loss of transcendence, the loss of cognitive worlds beyond the sociological. What is the social location of these ideas?" The concept of the "new class" is one way of responding to that question, he suggested.

288

In her introductory remarks Berger described her paper as an effort to provide a sociological basis for the discussion. "I'm trying to present to this group the social forces and processes at work in modern life and societies that have contributed to the making of the situation as it is, to present the broad outlines of the phenomenon of the new class and its processes. There exists today a class—or, if you wish, a category of people—that sees the task of guaranteeing a good life as its *raison d'être*, its property. Members of the new class defend that property with all their might. The term 'new class' refers to a group of people who are the producers and distributors of knowledge. They peddle non-material services. They apply symbolic knowledge to the existing world; they are concerned with interpersonal relationships and communication— like us. We are all part of the new-class phenomenon. It is a large class, it is growing, and it is very, very powerful."

Berger related the notion of the new class to the Marxist theory of social class, although she maintained that it transcends Marxist categories. In the new class it is cognition itself rather than economics which becomes stratified. The rise of intellectuals was an outgrowth of bourgeois capitalist society— related to science, technology, and capitalism—but not part of it. Berger was careful to distinguish the new class from scientists, who, as they matured, became problem-solvers. There is a fundamental difference in the mode of cognition between the old-class professions and the new-class professions, she pointed out. The old class is bound by natural science; the new class recognizes no such limits. Its way of thinking is abstract and highly subjective. This leads to an intellectual, rationalized "problematization" of the world. It removes the barriers between public and private; everything becomes public. The political arena is the scene of action. Feminism, Berger suggested, is an excellent example.

In opening the discussion, Neuhaus indicated that he regarded Berger's paper as extremely important. "What we have problematized as 'guaranteeing the good life'—we've done a little problematizing of our own here—is something we've done on the assumption that the issues which have arisen are not primarily the consequence of technological breakthroughs, whether with regard to abortion or euthanasia or reproductive rights or the family or whatever. Rather, we're assuming that something has happened in the way we *think* about these things; it's a cultural phenomenon.

"It seems to me very important to underscore Brigitte's point that the new-class phenomenon is something in which we are all embroiled. We're not talking about 'them.' This meeting, this conference, could not be more thoroughly new class. It is an enterprise involved in the meaning business—the manufacturing, managing, and marketing of meanings. And all of us make our living, in various ways, from telling people what life means. All of this means that we have to be credentialed by the very class we're worried about in order to get paid and to continue to do the things we enjoy doing, like holding conferences and writing books.

"Another point I'd like to underscore is that manufacturing, managing, and marketing meaning requires that you know more than other people. 'Knowledge is power' takes on a new and vicious force in this analysis of the class warfare in which we are embroiled. Since we constantly have to expand the subject matter, it assumes almost infinite plasticity. Everything is malleable, amenable to better understandings, which then can be managed in order to manipulate new self-understandings and new communities.

"The question is, Is Brigitte's ending inescapable? Some of us (and that might be most of us in this room) are, in a sense, 'traitors to our class.' That is, we're suspicious of the class of which we're a part. But are we not in fact—whether we like it or not—reinforcing that class? We are all training other people to accept our meanings. Therefore, are we not caught in a painful bind—that our dissent is something of little real force, since it is a tolerable dissent within a set of class interests?"

Neuhaus raised one additional issue. Prior to the rise of the new class, he said, it was the clergy, in addition to the intellectuals, who dealt with ideas and meanings. "What we witness today is that most of the professionally trained, educated clergy are as emphatically new class as any group: the social scientists, the psychiatrists—you name it. The clergy and the church no longer provide a fulcrum by which the new-class enterprise can be critiqued, criticized, and checked. Rather, they have been thoroughly co-opted by it, and have brought into the mix an aura of sanctity—the residual memory of transcendent beliefs—

which has reinforced very powerfully the unquestioned sacredness of the new class."

Distinguishing between Modernity and the New Class

Richard Stith of Valparaiso University, who spoke first as the discussion began, emphasized the urgency of the situation. "Any attempt to modify or critique new-class ideology ought to take into account its economic, political, and sociological bases, and how one might modify them," he said. "How can we provide for communities or classes or strata that would have some economic interest or other self-interest in formulating a different ideology? I think your whole project here, Pastor Neuhaus—of reconstituting mediating communities and creating places where there can be some realm of discourse—is along the lines I'm suggesting.

"On an international scale, how could we seek to remove some nations or cultures from the tremendous power of our Western new-class ideology? I think, for example, that it's very important for Iran to win its war. I said the same thing about Maoist China, which is now gone. But I think there are still a few places around the world where some alternative to new-class thinking really has hope of having significant economic and political power, and Iran is one of the last ones. I think we need to build up those pockets of resistance to the dominant ideological forces of this country."

Labeling this global reach as one of the best examples of the new-class mode of cognition, Berger replied, "It is most important to differentiate between modernization and the new class. The two things are not synonymous. Modernization, to my mind, cannot be turned around, but new-class influence can be moderated. What can be done is to identify forces that might give ordinary people, who are not part of a new-class mode of cognition, more power in the determination of their lives. There are very clearly public-policy issues at work, but most people who share your desire to derail the new class only reinforce it in the end, because they don't delineate sharply enough between modernity and the new class."

"Maybe the Third World has to go through the modernity process, and there's no way of escaping it," Neuhaus added. "But one would hate to think that people, in order to resist the new-class phenomenon that comes along with modernization, would have to be stuck in something like Mao's cultural revolution or the terrible tyranny of Khomeini." A brief discussion ensued over whether Mao and Khomeini should be thus rejected. Stith, by way of clarification, asserted that he was not advocating support of the Ayatollah Khomeini or Iran.

Moving into and out of the New Class

Noel Reynolds of the political science department of Brigham Young University then became the first of several to challenge the validity of the new-class analysis. "When you talk about class," he asked, "to what extent is it important that a class be identified in terms of family? I see the class you're describing as one that people can belong to at work, and then they can go home and become part of something very different. Many who would be identified as part of the new class have very strong footholds in the ordinary world. They are married to people who are very much in the ordinary world; their children get jobs in that other world; their parents are from that other world." Reynolds said he recognized the phenomenon being described but failed to see the "class coherence."

"You're saying these individuals live in compartmentalized worlds," Neuhaus clarified.

"Each of us lives in different worlds," Brigitte Berger added. "This is one of the great problems that modernization has opened up. We live in different institutional structures—ethnic, class, and religious structures—and different reference structures—the sports we like, our leisure-time activities, and so on. These are two different things, and this is one of the problems of the ordinary individual who participates in these different structures consecutively. Depending on how he thematizes whatever is important to him, he becomes his own enemy sometimes."

"But I think one can readily grant Noel's point that no class placement is all-inclusive of the totality of a person's life," Neuhaus interjected. "In the same town you might have a Chevrolet car dealer and a psychological counselor, and the counselor might fit the new-class picture much more than the business-class car dealer. Yet they might both belong to the All Saints Lutheran Church, which is a very important community in their lives. There they step outside of their class roles, which is part of the modern phenomenon—living in these different worlds that are in conflict with each other. This happens all the time."

"And with the family, too," Berger added.

"But in terms of the strength of class identity, is that class identity more or less tenuous in the new class than in others?" Reynolds asked.

The identity is very strong, Berger told him. "Everything you do—your politics, your religion, your leisure activities—is determined to a large degree by your class membership, whether you're aware of it or not."

Neuhaus continued, "When you have two people who encounter each other in a shared world—say, of religion—and one is business class and one is new class, the advantage that the new-class person has is that he or she claims to be able to explain—to provide a meaning for—the conflict that the business-class person experiences, whereas the business-class person does not presume to have such expertise. So that always gives the new-class person trump when it comes to the definition of the social reality and what ought to be done about it."

Berger offered an example. "I taught in a theological school one summer—I'll never do that again!—and lots of ministers came and gave examples of the kinds of problems that people brought to them in their parish life.

" 'What do you do?' I asked.

" 'I refer them to a psychologist; I refer them to a counselor,' they said.

" 'Has it ever occurred to you that they may come to you for prayer?' I inquired. And it had never occurred to them. In other words, these clergy had become new-class members, sitting at the center and referring people with problems to new-class consultations."

At this juncture Thomas Oden of the theological school of Drew University entered the discussion. "I think the new-class premise is tremendously valuable and promising," he said, "but I have some questions about it. Brigitte, you've described a kind of inevitability about new-class growth that I would like to see supported with either reasoning or data. It seems to me that this could be simply a speculative projection. This leads to a question about the newness of the new class. It does seem to me that we have many patterns—I'm thinking of the Platonic pattern of clergy in the medieval period, the clergy of the Reformation period, the scientific leadership of the eighteenth century, and so on—that anticipate what we're describing as the so-called new class."

Are We Part of the New Class?

Oden then raised a question that was to recur several times as members of the discussion group objected to their being included in the new class. "One more point has to do with whether we in this circle are new class or not," he continued. "I don't feel that we are. I don't feel that I'm a prototype of the new class, and I don't think this conversation is, either. One of the things that defines the new class is a lack of self-critical awareness. It seems to me that what we're up to here is a form of self-awareness that excludes us from that stereotype. I would like to say that I'm uncomfortable in my university and comfortable here precisely because I don't feel that I belong to the new class. I do in some ways, but I feel this is transcending it. I don't feel like a 'traitor to my class,' either. I don't feel like a traitor to the knowledge class. I feel that I am indeed committed to it in a self-critical way."

"But, Tom," Neuhaus interrupted, "why, objectively speaking, do you dispute the fact that you're in the new class? You're uneasy about the class of which you are a part, and you're self-critical and self-aware, but you are a member of the new class."

"No," Oden responded. "It seems to me that the way this term—the knowledge class or the new class—is being used, it is functioning as a kind of caricature. We are self-critical, and we're transcending that caricature."

"We're somehow missing something about the new class," Berger interjected. "The new class is very self-reflective. I have yet to meet a new-class professor who doesn't engage in constant self-reflection. But the difference is that they don't have a point of reference that is outside of the individual and the collective. This is where religion comes in. This is why I'm so insistent upon an understanding of religion. Religion isn't a psychological crutch; it doesn't come out of the needs of communal life. There must be a reference to the Total Otherness. I think that's why Professor Oden is uneasy about being identified as new class."

"Because he's a religious believer," said Neuhaus. "That's the source of his uneasiness."

"Of course. That makes the difference," Berger agreed. "The whole technological, scientific, new-class mode of cognition is indelibly imprinted upon the modern psyche. We're trying to demodernize, to jump out of the iron cage. But there is no jumping out."

Now Stanley Harakas of the Holy Cross School of Theology joined in, raising questions about definition and also resisting the new-class label. "New class rings a bell," he said. "There's something different about how people are functioning in our society. I think the terminology is useful. But you've got to find the touchstone for it." To suggest that the analysis of problems and the proposal of solutions is 'new class' is not particularly helpful, he continued. After all, these are things Saint Paul did, talking to the Greeks on Mars Hill. "These are functions of people who think and who address the human situation. If everybody who has done that is 'new class,' I don't think that helps. The statement was made here that everybody in this room is 'new class.' I'm wondering how useful that is. The inclusiveness of the term may create problems for its applicability and usefulness."

Harakas proposed that one of the things characteristic of the new class is the sense of wholeness, "the whole person, the *Whole Earth Catalog*—all that sense of totality. The trouble is in the parameters of that wholeness, which become restricted to this empirical world, which by definition rejects the transcendent."

Next he pointed to a second characteristic of the new class: the recognition of problems and attempts to solve those problems, "the development of knowledge or expertise (true or false) and the presenting of oneself as an expert in solving these problems, but only in terms of the negation of ultimate, transcendent reality." The term "spiritual" can't be used here, he said, because "this world" spiritualities fill the void—crystals, astrology, the idolatries of our age, which are substitutes for God.

Finally, Harakas suggested that the most powerful force conveying all this in Western society is "the popular acceptance of existentialist philosophy, which says, 'There is no meaning, but I must create my own meaning.' After the residue of a faith commitment has been washed out of the system, in its place has come a making of meanings."

Picking up the argument, Neuhaus pointed to the reduction of the language of religion and transcendence to the terms of psychological and collective needs, in the manner of nineteenth-century liberal Protestantism. "A large part of contemporary religion has basically bought into that proposition. Many theologians essentially agree with Feuerbach that references to God are human projections and are explicable by categories that are psychological and sociological in character. A person with that understanding of religion—even a clergyperson operating with that understanding of religion—would be emphatically a new-class person. What Father Harakas is saying is that someone who is operating with an authentic understanding of religion is in some sense not 'new class.' Stan Harakas is not involved in the manufacturing, managing, and marketing of meanings growing out of his expertise, but rather is involved in a community that is constituted by truth claims that are transcendent in character."

"Ultimate meanings," Harakas added. "The definition of new class is too inclusive if you simply say that the new class includes those dealing with meanings."

"But doesn't this get us back to the earlier objection of Noel Reynolds, to which the answer was that we all move in and out of these different roles?" Neuhaus asked. "When I'm standing in the pulpit of Immanuel Lutheran Church and preaching, I am emphatically *not* a member of the new class. While writing, I am socially located as a member of the new class. If we participate in the new class in this way, it doesn't mean that we have to buy into all of its assumptions. Indeed, we may be subversive of the new class; in that sense, the term 'traitors to our class' is apt. But we do move back and forth."

Berger sympathized with Harakas's refusal to be subsumed under the broad category of new class. "I also feel that I'm not 'new class' in much of my life. We get our frame of reference through the intentionality of what we're doing. I think Richard Neuhaus has put his finger on it by saying that we all move in and out of different worlds."

"But could Marx understand 'class' as something one jumps in and out of, almost as different functional roles?" Harakas asked. "I think the answer is no."

"I want to overcome the narrowness of the Marxist conceptualization," Berger replied.

Edward Gaffney of Loyola Law School of Los Angeles was not fully satisfied with this conclusion. "I agree with others, Brigitte: your definition of new class in the present version of your paper is far too undifferentiated for me to appreciate those moments when I'm participating in new-class thinking and those times when I'm resisting it," he said.

Structural and Cognitive Definitions of the New Class

Paul Vitz of the psychology department of New York University attempted to clarify with a simple cognitive definition of the new class: it believes "that there is no objective truth, that there are only self-defined meanings."

"I agree with that," Berger said.

Neuhaus added, "That's useful, Paul. You're making a distinction between functional and cognitive definitions, and functionally one may move back and forth, but that doesn't mean one is going to change one's mind about what truth is as one engages in the shuttle."

"But if I don't make that cognitive shift, that doesn't mean I'm a traitor to the new class," Vitz added, "because at that level I never was a member of that class."

"But someone looking at your role in society and what you do would say that you are emphatically a member of the new class," Neuhaus concluded, and Vitz agreed.

At this point Christopher Lasch of the history department of the University of Rochester highlighted an apparent difficulty in the concept of new class—lack of clarity about whether the term is being used sociologically or philosophically, with the correlative question about whether it is possible to move in and out of the new class. "If we're serious in talking about a class, it couldn't possibly be something one could move in and out of," he said. "But it's too simple to say it must be either a class or a cognition. It isn't one or the other; it's always both. Every class generates an ideology. So the distinction between a structural, social definition of the new class and a definition that stresses the way the new class thinks is misleading."

"That's what I try to argue," Berger interjected.

"I know, but that's where the difficulty comes in," Lasch continued. "One way of describing what I'm trying to advocate is a historical understanding of class. Classes are the products of certain kinds of historical changes. When we talk about the 'bourgeoisie' or the European nobility in an earlier period, those terms have historical content of a sort that 'new class' tends not to have. The ideas associated with them are at a high level of generality. We talk about 'new class' in terms of specific ideas. One has to ask, 'What are the historical changes that have allegedly produced this kind of class?' One would have to be quite specific about that, I think, and on the other hand, much more general about new-class ideas, the new-class view of the world. This conversation we're having is very confusing. Are the members of the new class to be defined principally as 'experts' as opposed to intellectuals? Are they heirs of the nineteenth-century bohemian, anti-bourgeois intellectuals? Are we talking about what people have referred to as the professional/managerial class? If that's the case, a lot of these ideological assertions won't stick, because they're not shared across the board. Your new class sounds like the last gasp of a dying breed of intellectuals, whereas the wave of the future is this new wave of managers and experts who don't care at all about these ideas."

Neuhaus said, "Kit, one point you make—which has come up in a number of different ways and which I think is a major contribution and achievement of the morning's discussion—is the very important distinction in these two ways of talking about the new class: one is social-structural, and the other is philosophical, ideological, cognitive."

"Ordinary People"

Now Jean Bethke Elshtain of the political science department of Vanderbilt University entered the discussion. "Brigitte, I have two concerns. The first is that your argument seems to rely, at least tacitly, on a thesis of 'false consciousness.' In your paper you talk about the engulfment of an 'unsuspecting population.' You talk about ordinary people becoming objects. It seems that you're casting your net in such a way that, to make your argument cohere, you're going to have to run with something like a thesis of false consciousness."

"Who is suffering from the false consciousness?" Neuhaus asked.

"Ordinary people—the unsuspecting populace, who are being engulfed, overtaken, and turned into objects by the new class. And presumably the new class is against the class interests of ordinary people. A part of this question is, Who are these ordinary people? What makes them ordinary? Is it the fact that they don't have college degrees? That they have incomes under $20,000 a year? Who are these ordinary folks? Aren't they riddled with false consciousness by not seeing that they're in some kind of war with the new-class practitioners who are out to turn them into objects?"

"You mean these common people think they're being served by the new class when in fact they're being exploited by the new class," Neuhaus clarified. He asked Berger if she would like to respond to Elshtain's question about false consciousness.

"That worries me a lot," Berger said, "because in the end I don't want to make out the new class to be the villain. The new class isn't making a conscious effort to lord it over ordinary people. At certain moments in our life we all become ordinary. The problem is that each of us is so many different things in one person. At certain moments I'm an ordinary person and at other moments I'm not. 'Ordinary' refers to the vicissitudes of life which we all encounter, which we try to approach through various ways of solving our problems—the scientific-technological mode, the traditional mode of common sense, or the new-class mode."

"When a sexologist, who is marvelously new class, has a fourteen-year-old daughter in high school, he or she moves very quickly back into the ordinary mode," Neuhaus noted, to some laughter.

Is the Concept Too Inclusive?

Jean Elshtain, who had focused the discussion on "ordinary people," went on to a second question. "Doesn't the notion of new class threaten to get so all-encompassing that it begins to lose its critical bite?" she asked Berger. "My example is drawn from your paper. You discuss over-medicalization as a new-class phenomenon, with people becoming less and less self-reliant and sure of their own resources in dealing with their own lives and bodies. So that's part of your thesis. But then, surprisingly, you draw Ivan

Illich into the discussion as yet another new-class person. Yet he is doing precisely the opposite of those encouraging over-medicalization; he challenges them. He wants ordinary people to get more control, to rely more on their own resources. My question is, How can you encompass both over-medicalization and the people who are trying to thwart that under one rubric?"

Berger responded, "Medicalization is not part of the new-class issue. Medicalization is basically old-line medicine. The new class does try to look at the human body in relationship to the larger universe, and it tries to rectify not the medical issues of the human body but the relationship to the larger structure. For that reason we now have epidemiology and preventive medicine. I was just part of a big project on the prevention of juvenile delinquency. It was proposed that if, at the age of eight months, a child exhibits some kind of behavior problem that can lead to delinquent behavior, that child should be taken away from his or her parents—'system deliveries,' that's what it is. The old-line doctor looks at the body as a puzzle and tries to explore everything. But in the end, this kind of doctor puts the options before the patient, and the patient has to make the decision. New-class medicine is of a very different type. It becomes a governmentally institutionalized program in which we all have to participate—sex education, teenage pregnancy prevention, what have you."

Some debate ensued over whether or not Berger should have used Illich as an illustration in her paper. Neuhaus interpreted the issue: "The new class typically says, 'You've got a problem you didn't realize you had. I'm an expert on your problem. I can explain it to you, and I can also give you meanings not only to help you understand but also to help you resolve that problem.' But others say quite the opposite of that, what in some sense Illich is saying: 'Look, trust yourself; follow your common sense.' Jean is saying that if you call that new class, isn't that including everything in the new class, so that it ends up losing its critical edge? Isn't what Illich is doing an anti–new-class thing, even though in terms of social location Illich is obviously a member of the new class?"

Here Richard Stith re-entered the conversation. "Brigitte, I agree with Jean in that I think you've misunderstood Illich. I think that results from a facile way of categorizing into only two camps: the old modern class and the new modern class. And you believe that since Illich is against the old modern class, he must be part of the new modern class. But there are still us weirdos around who are part of neither modern class, and I would claim Illich as a fellow weirdo!"

At this point Gilbert Meilaender of the religion department of Oberlin College interjected a thought: "I would like to put this out on the table, in the nexus between the last point Jean raised and the papers by Destro and Hauerwas. If the management of the environment and the use of medical techniques to guarantee the good life are important to the new class, is the direction in which Destro wants to go, which he sees as opposed to some legal goals, 'new class'? Destro is worried about developments that say, 'Let's back off from certain kinds of technologies.' He thinks we ought to go ahead and provide nutrition and hydration, and provide operations for infants under certain circumstances, and give certain kinds of care to senile people, and so on. Does that make him new class?"

Berger said she saw the problem. But Neuhaus answered, "No. Bob Destro wants very much to see an integrity of the legal subject matter as it defines itself, quite apart from any meaning imposed by the new class. And he says we're obligated to do those things supportive of the person—'person' being defined quite independently from our notions of infinite plasticity, what we can turn persons into, which would be the new-class way to go."

"But my problem is that the new class gets characterized in so many ways. I pick up one way, and you shoot the other one back in," Meilaender retorted. "It has been characterized in terms of management of the environment and use of techniques to control the environment and so on. It seems to me that Destro is calling for more of that in certain ways."

"So you're saying, like Jean, that maybe the new-class category is being used to explain too much," Neuhaus concluded.

Private, Public, and Governmental

Next James Burtchaell of the theology department of the University of Notre Dame entered the discussion, suggesting that there are professional parallels between the old class and the new, and pointing to the diffusion of new-class modes of cognition into the old class. The old class has its arm just as deeply into government and into public persuasion as the new class does. "Brigitte made the point," he went on, "that the new class is to be characterized by its determination to bring to governmental resolution virtually all the large projects of social policy that it wants for the protection of all. But if you went down to Capitol Hill, you'd probably find that representatives are spending as many hours listening to unions and the Chamber of Commerce and representatives of the rubber industry as to promoters of new-class programs."

Burtchaell went on to question several references that Berger made in her paper to religion and the church. Citing one of Berger's assertions that the new class has problematized and publicized certain concerns (such as programs relating to children, women, and moral education) that were previously thought to be private, Burtchaell noted, "They seem to me to be concerns that were those of the family and of the church, and those aren't private matters. The family wasn't private, and the church certainly wasn't private. The dissociation of public and private there was something that I think might have been reconsidered."

Berger conceded the legitimacy of the point, although asserting that these were not concerns of the church in earlier years. But Neuhaus added, "It was outside the purview of 'the public' understood as governmental, however."

"An understanding that I would wrathfully oppose," Burtchaell asserted.

"Yes, but understanding 'public' to mean 'governmental' certainly is a typical and deeply entrenched new-class view," said Neuhaus. "So perhaps Brigitte would have been better advised to talk not about making these things public but about the governmentalizing of these issues."

"I'm not sure I would go that far," Berger responded. To Burtchaell she said, "I think there's a deep difference between us. There are a number of things that modern society has done over against traditional society which I cherish. One of them is to guarantee certain private spheres, which are indeed private."

"Spheres beyond governmental control," Neuhaus interjected. After an exchange with Berger regarding the public nature of the church, he continued, "Maybe what we should pin down here is that an important challenge has been raised with regard to the public-private distinction, and I'm suggesting that what's missing in the paper is the distinction between public and governmental—and political in the sense of governmental."

Jean Elshtain re-entered the discussion briefly. "I just wanted to flip over Brigitte's point and suggest that claims to privatization can be a vicious form of politicization—especially in the realm of sexuality, where we politicize by privatizing, by saying, 'This is nobody's business but my own.'"

Berger agreed, and Neuhaus added, "In our kind of legal and political system, where individual rights are trump, so to speak, the claim to privatize can be a politically and legally powerful thing, and can paradoxically lead to greater governmental action."

The Vulnerability of the New Class

Is the ideology of the new class, particularly with regard to ideas surrounding current disputes over the good life, sure to prevail? Not necessarily, Paul Vitz suggested.

"I'm a psychologist and I do counseling," he said. "And I refer my clients, finally, to a spiritual advisor. As a Christian and Catholic counselor, my patron saint is John the Baptist. I have to 'make straight the way for the Lord.' This means that when my job is finished, my clients go on to a spiritual advisor. This experience that I'm having with clients is, I think, valid. I think the fundamental vulnerability of the new class is that it's not based on truth.

"The new class assumes that there is no fundamental truth with respect to human life and meaning. The history of experience and wisdom in the world as we look at it suggests that new-class members are the sophists, as has been implied. And therefore I think the new class is extremely vulnerable, because it is increasingly anarchic intellectually. It had a much more common understanding of its function twenty or thirty years ago with a certain kind of positivistic psychology, for example. Now psychology is splintered into a thousand different fields. And the movement is clearly *toward* religion, not away from it, almost universally. Even the American Psychological Association is beginning to have large groups attend its sessions on religion. So I would say that the new class does have a great vulnerability in its intellectual anarchy, chaos, and insecurity. The truth is not with the new class.

"The evidence is clear: the best that psychology can do for a person is to bring about a modest positive change, and there's no evidence that the modest positive change is related to the ideas of the psychologist. It's probably related to these facts: that the person has made a commitment to therapy, that the person is spending time on the problem and paying money for it, and that the psychologist has the capacity to support the person, help the person, and—if he or she has a certain degree of wisdom—illuminate the situation. These things have nothing to do with the actual course content with which we educate our clinical psychologists. And this absence of fundamental validity in many of the new-class concepts is finally beginning to seep into the culture at large. It will undermine what cultural legitimacy there is."

Vitz went on to make several additional points, his second one being that the new class is economically vulnerable, since its members are paid indirectly for their labor. The government's ability to continue to pay for everything is limited, he noted. He suggested that the new class is like the old aristocrats, with taxpayers as "serfs," paying the bills. "There is an exploitative economic component here," he said, which will become politically polarizing.

Third, Vitz said, the new class is institutionalized primarily in the universities, which make up the "seminary" for this class. The university is the "filter system" that provides the governing members for our society. Some of the things to discuss, he pointed out, are the weakness of the university, how to address corrections within it (since there remains a residual commitment to the truth), and how to provide a framework within which a challenge to the new class can take place.

"Finally," he said, "as intellectuals—and even as new-class members—the best we can do is to provide ideas. We don't change history. But I don't think we should overlook the possibility of a serious crisis for the new class and for the West itself. The Islamic revolution is something the new class never predicted and couldn't expect. It is, in my judgment, the anti-revolution, and it may have as much power in its long-term impact as the French Revolution. There could be a radical change in the assumptions that govern modernism and the new-class ideology. The new class is more vulnerable than we think. But the most we can do is to analyze it, to provide intelligent understandings of why it is wrong and what would be the new way to move, and then history will be in the driver's seat."

Jonathan Imber of the sociology department of Wellesley College returned to the question of whether one is in or outside of the new-class mode of cognition. New-class theory suggests that all people amount to is their roles, he said. The greatest formulator of this position was Erving Goffman, who brought a special skepticism to what it means to say that one has an identity. "Suppose I declare that I can't get pregnant. According to the older orders of authority, which stipulate my maleness, such a statement is absurd. Of course I can't get pregnant. But the new-class mode of cognition challenges this. It says, 'Maybe you can. Let's propose surgery. Let's propose other options.'"

"This gets us back to the plasticity of the self," Neuhaus added.

"It seems to me," Imber put in, "that it was the historic obligation of the churches and also of the family to secure the kinds of certainty, the kinds of identity, and the expectations that accompany certainty and identity." He referred to the last sentence of the first draft of Dr. Berger's paper: " 'It is a much more difficult, much more personal achievement to express coherently a faith in values that deviate from those currently monopolizing American culture.' "

"I read that," Imber continued, "to mean that one says, 'I stand for this' or 'I can't stand for that.' This perhaps gets us back to a characterization of the liberal ethos, the inability of people to say no, to recognize what limits are about."

Neuhaus commented on this very valuable contribution to the discussion. "We have perhaps drifted far from the set of concerns that initiated these meetings—'guaranteeing the good life' and all the bizarre things that are being proposed in fetal experimentation and abortion and euthanasia and whatnot—but in fact at the heart of this whole discussion of these questions is the capacity to take a stand. And why it is so difficult in our culture for someone to say some things are unthinkable and should remain unthinkable."

Berger expressed appreciation for all the suggestions about her paper, and brought the session to a close with a summary. " 'New class' doesn't explain everything," she said. "It's simply a useful way of looking at things as they have come to be. I do think it is a class; all the data show a connection between a certain mode of cognition and social structure and location. The difference here is really between a style of cognition and the content of cognition. What is really at issue is a *style* of cognition, not specific ideas. No one owns new-class ideas. They fall across the political spectrum. They go into every little hamlet; they're part of modern life."

Here Neuhaus interrupted. "Brigitte, what you just said violates what I understand to be the empirical data that you referred to—namely, that content (that is, the specific judgments with regard to very specific issues, most specifically with regard to the issues this conference is concerned about) does indeed fit into new-class theory."

"The point I would make very emphatically is that a certain mode of cognition leads to certain ideas," Berger responded. "That's the issue. I do think the Marxist concepts are very useful. While this class of people does meet all the Marxist requirements for a class, it is more than that. It is a much broader class, and that's what my conclusion is."

"But as Kit Lasch and others have pointed out, Marx wouldn't have people moving in and out of a class," Neuhaus objected.

"But for that reason," Berger replied, "I think the Marxist notion is much too limited. I do think this is a major contribution to our philosophical understanding of what's going on.

"Another point that is equally important is the notion that the new class is not synonymous with modernity and modern society. To be critical of the new class—and to some degree I am critical of it, though not in every way—does not mean one equates modernity with new-class behavior and ideas. Modernity as it has developed in Western culture has a lot to show for itself. It has brought to the fore individualization, which is part and parcel of modern life, and which we do not want to be subsumed in the Ayatollah's Islamic republic. The West has something to stand up for, and in our animosity against the new class we shouldn't quickly throw out all the achievements of Western culture.

"But these dynamics at work, which have culturally and cognitively assisted the rise of the new class, have taken a new turn with the rupture of individuals from their religious embeddedness. Thus, again, I come back to the religious question. Religion is culturally necessary and important, but it is much more important in terms of human existence. This is a question for the theologians to work out. But I know many theologians today, and I don't trust them. So in the end we 'vulgar' religious people have to do the job ourselves—that's what I mean by 'ordinary people.' "

GOD, MEDICINE, AND PROBLEMS OF EVIL

Central to the new-class enterprise is the problem of suffering, since the elimination of suffering is the rationale behind many attempts to guarantee a good life through technological intervention. Suffering, therefore, was the focal point of the second session of the conference.

Stanley Hauerwas of the Divinity School of Duke University introduced his paper by discussing the problem of theodicy, which, he said, was not an issue for early Christianity but has been an issue since the early seventeenth century. It comes out of a kind of thinking which assumes that theology starts with the doctrine of God in a philosophical framework, apart from the soteriological framework, the only framework within which the Christian doctrine of God is intelligible. The so-called problem of theodicy was created once a theoretical problem about God's existence had been created, and the problem of suffering was posed from a universal perspective.

For the early Christians, some kinds of suffering (such as persecution) were inherent in baptism; other kinds (such as the death of a child) could be made meaningful by faith. Nothing within the Christian tradition suggested that illness should be received without seeking a cure. Suffering, however, was seen not as raising a question about God's existence but as presenting the practical question of how Christians should care for one another. According to Hauerwas, the book of Job has been placed within the hermeneutics of the Enlightenment and thus has been misread for centuries.

Nor is theodicy a problem for most people today, Hauerwas continued, given the anthropocentricity of modern culture. Indeed, the problem is "anthropodicy," for if the assumption is that humankind now determines the meaning of our existence, then we ought to have the power to remove those impediments that make the human project less possible. We don't tend to think hurricanes raise the problem of evil, but we think sickness does.

Why? Because we think we have the technological means to alleviate if not eliminate sickness. And if we cannot eliminate an illness (such as Down's syndrome), we end up eliminating its victims. Medicalizing our problems avoids facing those limits which are not only limits but also possibilities of service to one another. Christians have a narrative that makes illness—including the hardest case, the illness of children—bearable.

In opening discussion about the Hauerwas paper, Neuhaus pointed to the modern perspective on suffering as a "problem" that must be "solved." He noted that in the critical responses to his *Commentary* article on eugenics (included herein), readers basically asked, Why is this alarming, since all of these developments (treating infertility, using fetal parts, etc.) are intended to relieve suffering? He noted that the group in California that had pushed unsuccessfully to put euthanasia on the ballot is called Americans Against Human Suffering. Their literature implies that they are against suffering humans, he pointed out. Suffering humans are a violation of the meanings we would impose, and think we can impose, with the technology to enforce it.

Neuhaus tied the problem to Brigitte Berger's discussion of the new class. Atheism, he said, begins with the Enlightenment construct of God rather than the alternative Jewish and Christian faith in a God known through suffering. There is something in both the ideological and the social-structural dimensions of the new-class phenomenon that requires "drawing in the boundaries of reality" to end up with a kind of practical and functional atheism. This not only makes it impermissible to admit the biblical account of God; it also means that in the realm of devising answers, there will always be a better answer, an answer that will solve our problem of suffering.

Suffering as an Intellectual Problem

As the general discussion opened, Noel Reynolds said that he thought Hauerwas's idea that the problem of theodicy derives from its place in philosophical theology was an important insight. "But I think it may be a little too easy to get rid of the theodicy problem and its challenges to traditional belief, and I wonder if you haven't stepped too quickly in that direction. In all cultures, one of the main things that causes people to question their received worldview is suffering. Very few worldviews really enable people to deal with suffering when they experience it directly. It's not only a matter of how we respond to suffering and how we treat each other when we're suffering, but also a matter of how we understand it."

Stanley Harakas also questioned Hauerwas's rejection of theodicy. "People raise the question," he pointed out. "It won't work to walk away from it and to say, 'You're asking the wrong question; let me tell you what the right question is so I can answer that one.' You must somehow respond. People bring their whole world with them when they ask questions about suffering. Suffering should be alleviated wherever possible. That's a command of Christian love, and that means medicine should address it. But in this finite world we can never totally eradicate suffering, and we must also teach people how to meet the suffering that cannot be eradicated."

Here Thomas Oden joined the discussion. He defined suffering as "the human condition of the conflict between nature and self-transcendence." We are beings who by our very nature are suffering. And wisdom is the discovery of our human condition.

"There is great wisdom in the classical Christian tradition," Oden said, questioning why Hauerwas was so opposed to the speculative, the abstract, and the theoretical over against the practical. "Both the Jewish and the Christian communities have been greatly concerned with ministering to and caring for people in the midst of suffering. That in turn has elicited a long tradition of reflection, which on the whole has been called theodicy. There is an entire mode of reasoning concerning the way in which God brings good out of evil, and the interfacing of providence and pedagogy whereby we are to learn from every human experience. I've learned a lot from your paper, Stan, but you haven't convinced me that suffering is fundamentally a Christian problem. It seems to me it is a human problem that emerges out of the human condition."

Hillel Fradkin of the Bradley Foundation continued to examine the theme: "Strictly speaking, theodicy is a discussion of God's justice. But the biblical revelation in the Old and the New Testaments seems to be based on the premise that there is no such thing as God's justice. Whatever God's relations are to humankind, they are not relations of justice or injustice. Justice falls to the sphere of human activity, human relations.

"On that basis it seems fair to ask, 'Why is there a theodicy?' You wonder whether that is the proper way of putting things all together. But however much theodicy may have been transformed by the Enlightenment, it wasn't invented by it. Suffering can't really be regarded as a gift of God or a function of what God is trying to do to us, and it also can't be regarded as something that is really a result of human injustice. It's just kind of there. The immediate response is, 'Who's going to pick up that ball? You can't just leave it on the table.' But that isn't a reasonable response. If the responsibility isn't going to be assigned to God, then human beings are the only others who can take responsibility for it. And then the question is, On what basis do they take responsibility? So it seems to me hardly unnatural that, the influence of the Enlightenment aside, people try to break out of that bind."

Gilbert Meilaender also joined those who questioned the elimination of "the intellectual problem." He began by referring to Hauerwas's treatment of the book of Job. "There's more in the biblical literature than you take account of. Even if you find another way to read the book of Job, have a look at Habakkuk and Malachi, for instance. The question there is not simply 'How could the electing God do such a thing to his elect people?' It's a more general kind of question. It's hard to eliminate it in that literature.

"Second, I'm not so sure the intellectual problem stems simply from some kind of Enlightenment notion of an abstract God characterized as perfect in a negative way and so on. It seems to me that the problem is generated even more strongly if we begin to describe that God in the ways the Christian narrative does—as caring for us. Precisely to the degree that we think we have good reasons for thinking this God actually has our well-being at heart, there will be occasions when this problem will arise.

"And on those occasions—this is the third point—what we're looking for in part is some kind of account or story or whatever that makes suffering not only bearable, to use your term, but also intelligible. And seeking that is not a way to overcome it or get rid of it. It falls under the rubric of loving God with the mind."

Hauerwas responded. "Real people ask the question about suffering, and a response is called for. But the problem is that oftentimes people have been educated to ask the question out of a tradition that hasn't been informed by Christian presuppositions. The question is asked out of the presumption that it is basically possible to eliminate illness and that 'if my child got polio before there was a cure, a deep injustice has been done to me.' I think regular people ask the question that way."

Christopher Lasch suggested that if the question is asked in that form, it may be preferable simply to deal with it out of the resources of the Christian community rather than explaining that it is not really an intellectual problem to be solved but one to be lived with on the basis of the community's resources.

Next Hadley Arkes of Amherst College seconded the suggestions of Fradkin and Meilaender that there may not be anything in the Hauerwas project that needs to put him at odds with the Enlightenment or with the notion of moral autonomy (though it does put him at odds with those misuses of autonomy that he correctly mentioned). "But at the same time," Arkes noted, "my inclination is to say, 'Go for it!' It's like Lincoln's story about the wife seeing her husband at war with a bear. She's not sure what side to take, so she says, 'Go, husband! Go, bear!' Similarly, 'Go for it, Stanley! Go for it, Enlightenment!' "

Useful Suffering, False Suffering, and Masochism

Sidney Callahan of the psychology department of Mercy College agreed with Hauerwas that the practical problem of what to do about suffering is more important than what to say about suffering. "I don't think Christianity explains it, but it sure gives some ways to deal with it." However, she went on, there is a universal human nature. "That part of the Enlightenment was right. We all know what suffering is. All cultures have an idea of the Fall."

"Catholics sometimes deal with suffering," she continued, "on the basis of the idea that one's suffering 'fills up' the suffering of Christ, that one's suffering can be used vicariously, that one can 'offer up' suffering and thus that one's suffering is never meaningless, that it will be used in some way in the 'treasury of grace.' Some modern Catholic theologians cast doubt on that idea, and I've said, 'Oh boy, are they wrong!' "

Neuhaus added, "A very powerful cultural demonstration of that occurred when John XXIII died. It went on for about fifteen days, as he died of cancer, and each day he gave a little homily on what sufferings of the world he was offering up his own suffering for. For a lot of people that was an introduction to a concept of suffering and death which was quite novel."

Callahan continued by noting that among ordinary people, far from the new class, this is a powerful idea. But it can encourage people to glorify suffering—to engage in a kind of masochism. "It's only *necessary* suffering that can be offered up" she pointed out. "It was when Thomas More was driven to the wall that he became a martyr. Relief of suffering is also part of the Christian tradition: healing, 'every tear will be wiped away,' and so on. So it seems to me that how to strike that balance is important."

Paul Vitz commented on Callahan's warning about the masochistic glorifying of suffering, which he referred to as "the notion of false suffering." He added, "There are other kinds of what I would call false suffering. Some of the modern concern for suffering is itself a kind of false suffering." He pointed to the adult response to the suffering of children. "There is a sense in which what bothers us when we see a child in pain is not what's happening to the child but what that child is making happen to us. The child can represent an extension of our wants and dreams, and for the child to suffer is for our own projected narcissism to be attacked."

There are also situations, Vitz said, in which we "grieve with alacrity" as a way of pointing out the horribleness of the people (our enemies) who are causing the suffering. "Some people who dislike Israel are delighted with the recent repression of the Arabs, because it gives them a way to attack someone they dislike. Some people with grudges against God use the suffering of others as a club to get back at God. I'm not saying this is all of suffering, but there is a way in which suffering can be false when it's used for another purpose."

"But it's not false because it's unreal," Neuhaus clarified. "If I understand you correctly, it's false because the person falsely understands the nature of the suffering."

"Yes," Vitz confirmed, "and false in the sense that it has no weight of argument in the theological context. It is not a critique of God. It's the kind of suffering that can be created by our need to control things and our need to have perfect children. That kind of thing is closer to narcissism and can be psychologically explained, unlike the other fundamental, common suffering that humanity has been going through since the beginning."

Jean Elshtain described a situation in which children dying of incurable cancer were not told by their parents that they were dying. But the children knew, and they created their own narratives to make sense of what they were going through. In their narratives the children said, "I will keep up the pretense that will help my parents feel better about the fact that I'm dying."

"Suffering," Elshtain went on to say, "is always a sign that the culture reads. When people are suffering terribly in ways that are visible to others, they cross a boundary into another realm. We need to think more widely about how our particular world deals with the question of difference, of the other. And this ties back into the question of the new eugenics, which is a way to attempt to eliminate our problem with otherness or difference by eliminating the different. We get rid of Down's syndrome by disallowing any Down's syndrome children to be born. This becomes the way in which we handle that very complicated problem of which visible, prolonged suffering is one category—people who are 'other than,' 'different from.'"

Hauerwas spoke to the question of eliminating the mentally handicapped because they are suffering. "As a matter of fact, they aren't suffering by being mentally handicapped," he said, "but what we always do is attribute a kind of suffering to them in terms of how we would feel if we were mentally handicapped."

Neuhaus pressed him on this point. "So, Stan, you're saying this to the person who says that this Down's syndrome or severely handicapped baby would have a very hard life, and that if the prohibition against taking innocent life can be overridden, suffering is a reason to override it. Your response is, 'That life would be a hard life only by your definition of what would be hard.'"

"Let's not romanticize the 'hard life,'" Hauerwas replied. "That's often an argument for sustaining the lives of people who are profoundly mentally handicapped: 'I know this family that became an exemplar to us all because it had this profoundly handicapped child.' That argument is terribly destructive to the handicapped child. The worst thing you can do to people is to train them to accept the language that says they're victims."

Both Neuhaus and James Burtchaell asked why the exemplary care and human development of the family is not a legitimate secondary reason (in addition to the the primary reason of our stewardship of life) for preserving the life of the profoundly handicapped child.

Hauerwas related this question to Thomas Oden's point about the relation between providence and pedagogy, and Sidney Callahan's point about how suffering contributes to the "treasury of grace." "My problem with that is that it can be an invitation to masochism. I may have a nagging mother-in-law or a tough marriage, but those don't involve the kind of suffering that God calls me to, which is Christ's cross. Christ's cross is a particular kind of suffering one should expect when one confronts the powers that deny God's reality. It's not general suffering."

Neuhaus demurred about using the term "masochism." "Today Saint Paul's talk about rejoicing and exulting in filling up the sufferings of Christ and so on would clearly fit, by psychological criteria, the category of masochism. In fact, even the cross of Christ and the death of our Lord himself would be included. This, I think, highlights the radicality of the conflict between the biblical proposition and culturally prevailing notions of suffering and masochism."

Stephen Post of the Center for Biomedical Ethics at Case Western Reserve University added that while it is true that there are some accounts of families which develop tremendous moral idealism when confronted with a suffering, handicapped newborn, "there's another body of literature which is much more suspicious of the experience. We can't make any grandiose generalizations about what kind of active love is brought into the world by suffering. There's something banal and platitudinous about that, and it troubles me."

Finding the Meaning of Suffering

At this juncture Myriam Puig Abuli, a pediatrician from Caracas, Venezuela, entered the discussion. "Suffering is a human condition," she said. "Doctors find out every day that it can't be eliminated, and to think that it can be eliminated is to give medicine an omnipotence it doesn't have. I would say the question is, Does suffering have meaning? You see that in medical practice. When you have to tell parents that their child is going to die, or that their child is malformed or has some type of handicap, you find that there are two kinds of people. Some people will find meaning in suffering; others will not. When you have parents who see that meaning, you know that the child is 'safe' in a sense, and will have a future."

"The question is," Puig continued, "how do we present the situation? Is suffering the end of the world, or is it only the beginning? If I see only the 'perfect,' then in the short run suffering will be the end of the world. But if I am able to give these people a sense of transcendence so they can go beyond themselves, they will find meaning in suffering, perhaps a reason for it. The search for meaning is one of the most important problems of life. And sometimes one faces a situation like a child's handicap or terminal illness, and the suffering involved can provide an opportunity to find that meaning. I think the question is, How can we help people face that situation?"

Puig talked of the suffering involved in motherhood, which is met by transcending the self. "When you are not able to go beyond yourself," she said, "even the smallest inconveniences are understood as causes of suffering." She spoke of the way in which, in an earlier period, people learned the "sick role" from someone in their household who was dying. Now, in the absence of such role models, we don't know how to be sick; we don't understand sickness and suffering, and we expect more of medicine than medicine can deliver.

Next Stephen Post spoke, expressing appreciation for Hauerwas's essay but offering a critical comment. "When I hear that suffering is supposed to be meaningful to me at least in part because it gives other people an opportunity to demonstrate active love, that leaves me wary. It makes me feel like I'm a means to the end of the caring community. I think, Stan, that you've tried to mitigate the personal agony of suffering by diffusing it into the goodness of the community. But even if that were adequate, it's not of wide public value, not in the argument against active euthanasia or mercy killing. The real issue is whether we want to make an exception to the prohibition against taking human life. I think not, but I can't then appeal to your kind of contextual-communal argument based on narrative and the like."

"Steve, how did you make that move?" Neuhaus asked.

Post replied, "I think what we're saying here is that we want to avoid active killing, and one classical argument in that direction is that suffering is redemptive, that there can be ways of understanding suffering which will make us more willing to endure it so that death can take its natural course. I don't think that carries much weight in our modern world, and I'd rather make the argument on the fundamental prohibition against taking innocent life."

Now Edward Gaffney turned to the way in which Hauerwas addressed suffering within the context of the Christian community. "The absence of specific reference by you, a Christian theologian, to Jewish experience is a major lacuna" in the paper, he said. He expressed appreciation for the references that Hauerwas made to Old Testament sources in his oral presentation of the paper. Hauerwas commented that his interpretation of those texts was distinctively Christian, and not one that a religious Jew could accept.

Gaffney continued, "It's appropriate for Hebrews to speak from the Hebraic faith, and for Christians to be respectful of the Jews' experience of suffering in a redemptive mode. For instance, response to the Holocaust is something that's appropriately Jewish. It's not that I want you, Stan, to be the articulator of that, but surely it brings a cleansing of vision. Surely the ideas of Mohandas Gandhi, in the face of political repression in India, involve a cleansing beyond the Hindu vision."

Neuhaus asked Gaffney if he was trying to push Hauerwas in a more universalistic direction. An extended discussion ensued over the meaning of suffering from the Christian perspective, and whether the greatest suffering may be caused by those things which, superficially, seem to give the greatest pleasure. James Burtchaell maintained that the greatest suffering is "not being in charge. Accordingly, lots of people don't know suffering for what it is," he said. "Many of the things we think we're suffering are in fact trivial things. The big sufferings—like being a Christian, for example—are too big even to look at." Conferees questioned this understanding of suffering.

Stanley Harakas said, "I suppose that could be translated in terms of the human condition, understood as a result of broken relationship with God." But this isn't the normal use of the term "suffering," he went on to say. "The normal use of the term has to do with the kind of disorderedness that comes to us—but how far can we take that? If a child gets hit by a car, I don't think anyone has a problem with understanding what that means in terms of suffering. But are there limits to what we may define as suffering? Am I suffering if my nose is too long?"

Burtchaell responded. "In our consideration of the good life, we've kept coming back to the hubristic tendency to have it the way we want it. And often that has had very little to do with a right sense of order. We define the good life by will rather than by discernment or true need. I'm trying to argue that when we talk about suffering, we make a pygmy out of the issue by looking at only one type of suffering."

At this point Neuhaus put the recent exchanges in context. "We're interpreting suffering within an increasingly explicit religious framework, which is fine, but the structure of the conversation seems to be moving us toward saying that this understanding, which many—perhaps most—of us here would accept, is simply so countercultural, so out of sync with prevailing cultural options, that the answer to our original question is that the only way our societal situation is going to change is through a great religious awakening."

Here Brigitte Berger broke in. "I wouldn't advocate that line or any line like that. I hate to be a party pooper among professional theologians who are engaged in exquisite, sophisticated discussions on the meaning of suffering! Let me start again with Stan's paper, and the model of the mentally handicapped child."

Identifying herself as the mother of a handicapped child, Berger went on to say, "I understood many of the things Stan is saying in his paper, and many of the thoughts he articulates have come to my mind as well. Different cultures have different repertoires to deal with human suffering. The Hindu culture of India comes to my mind, and the Hindu notion of not interfering in human suffering. There's a whole theology to that. I doubt if anyone in this room would subscribe to this kind of theological argumentation. The Judeo-Christian tradition doesn't permit us to do that. We believe that this life, here and now, is our individual life. We feel an obligation to live this individual life and an obligation to our fellow human beings to interfere in human suffering, to alleviate it; that's the Christian tradition.

It is out of this kind of combination of factors that modern science came into existence. This is what brought us medicine, science, and the promise of 'guaranteeing the good life.' The irony is that what gave rise to modern science no longer undergirds it. To be sure, religion tries to give meaning to suffering, but in a secular society most people no longer have any language for trying to do that. And perhaps what everyone has been saying here is correct—that human suffering serves a great variety of functions. Perhaps there's a redemptive role, perhaps there's a paradigmatic role for suffering, and I'm willing to accept this. But I don't want the argument for the protection of the handicapped made in these terms. I don't think the handicapped are well served by them."

Medicine and the Goal of Caring for the Suffering

Berger then went on to discuss the failure of modern medicine to accept its own limitations. Neuhaus pressed her to follow through on her own argument for the protection of the handicapped. "Why take care of the handicapped?" he asked. "What are your reasons?"

"If there's a small thing that can be done which would fundamentally add to life, I would be foolish not to do that," Berger responded. With regard to her own son, she said, "He is God-given to me. But if something could be done to improve his condition, it would be wrong not to do it." With regard to the experts, she said, "The funny thing is how little we know, and yet we do know more than we knew twenty years ago. We have more knowledge about what works and what doesn't."

Burtchaell asked her, "When you were working hard taking care of your son, and yet that wasn't working, were you failing or succeeding?" Discussion followed about how success or failure is defined, and Berger commented on the importance of acknowledging the limits of what medicine can do. Many people want to deny the existence of such limits, she pointed out. "Part of our cultural heritage is the idea that if only we find the right formula, if only we find the right method, then something will happen. The new class has been co-opted into that kind of cultural understanding of human life."

Neuhaus sought to clarify Berger's point: "So 'caring is problem-solving' is basically the paradigm the new class is working with. Its members didn't want to know the reality, the near-certainty that the problem wasn't going to be solved because they wanted to keep believing that they were engaged in problem-solving. Is that what you're saying?"

Berger wasn't totally satisfied with this. "I want to make a case for modern medicine," she said. "I don't want to throw the whole thing out." Discussion continued, and participants generally agreed that success or failure isn't determined by whether or not cure takes place. There must be a goal beyond cure.

Edward Gaffney brought up the Alcoholics Anonymous program, in which there is no assumption that alcoholism is something which can be cured. Hope for recovery, he said, lies in acknowledging one's powerlessness over what is happening. "But from that springs an extraordinarily rich spirituality."

Neuhaus asked, "If there were a cure for alcoholism—a shot you could take—would you wish for that?"

"I'm not trying to launch a diatribe against medication," Gaffney answered. "I'm merely describing what's known. Certainly no cure is in sight, and right now I can only describe the rich spiritual benefits of accepting the perilousness of one's condition."

Burtchaell agreed. "What one comes to is not a restoration of what one would have been before all this. But it is immeasurably greater than what one could have come to if it hadn't been for all this."

"Alcoholism is a crucial example," Hauerwas noted, "because while we know there are genetic and chemical bases for alcoholism, it is also true that the only way an alcoholic gets better is to take responsibility for what is not in his or her power."

Now Stanley Harakas spoke, expressing discomfort initially with what he perceived to be an implication of Hauerwas's perspective that healing is less than fully Christian, and that in the face of suffering, comforting is the major responsibility of medicine.

Neuhaus pointed out the question being raised was about the definition of medicine. He recalled Berger's earlier suggestion that medicine was at its best when it had a very unpretentious kind of technical understanding of its mission defined by the human body, and when it recognized that it could do certain things to heal disease, help the healing process, and relieve pain. Neuhaus said that this modest definition can be perceived not only as a work of love but also as a restraint against the more vaulting definitions of medicine that bestow overarching meaning on life in its new-class manifestations.

Next came an exchange with Hauerwas over the extent to which such things as infertility or any experienced unhappiness or frustration are to be understood as suffering and addressed by medicine.

Medicine and the Meaning of Suffering

Now Jonathan Imber entered the conversation. After discussing the culture's traditional enthusiasm about medicine, the expectation of uniformity of medical treatment but the reality of varying levels of such treatment, and the problems this brings in a democratic society, he turned to the question of suffering. "Different faith traditions have given very different kinds of answers to the problems of suffering," he said. "It's not a question of expecting the government to decide how those answers will be given. It's a question of seeing that the polity depends on the fact that there are clear but different voices suggesting ways to come to terms with suffering."

Referring to Berger's earlier comment on the absence of a language to deal with suffering, Robert Destro of the Columbus School of Law of the Catholic University of America suggested that perhaps the reason we can't explain suffering is that suffering is part of the human condition, and we can't explain our own condition. "When we discuss the handicapped child, or the child with AIDS, for example, it's difficult because what we can't explain is not so much the child's suffering but the reason for the child's being there in the first place. It's that lack of an explanatory language that's really confounding us." Consequently, we turn to scientists, he went on, following up on a point made by Myriam Puig. "We think we can finesse the question of why we're here if we can just find the scientific solutions to our problems." He found an example in the recent testimony of AIDS victims before the U.S. Civil Rights Commission. The first group that testified felt betrayed: "It said, 'We were autonomous, and now we have to recognize that we are dependent. You promised us better living through chemistry, and chemistry didn't deliver.'" But the last witness talked only about what he would do with the time he had left. So, Destro concluded, "when you go back to the human subject at the center, it comes down to, 'What is the value of that subject, and why is it here?'"

Neuhaus interpreted here: "Bob, your point is a very valuable one—that is, why is it that we think suffering is strange? Why is it that we think it raises questions? I think it was Tom Oden who first made the point that suffering is inherent in the human condition, and others have said the same thing in different ways. And you say, very suggestively, in the case of AIDS victims, 'Why didn't they find their very existence the strange thing to wonder about rather than the fact that they suffer?' "

Richard Stith, referring to Hinduism, said that Krishna's teaching of non-attachment is very close to what we're aiming at here—"the notion of pursuing remedies for suffering but not being attached to the results. What matters is not the result but the pursuit, the carrying out of one's duty of caring. And part of the caring is trying to bring about the result but not caring whether or not that result occurs. The Hindu teaching is radical; one isn't supposed to care about success. Also, in the Christian tradition we're not supposed to care about failure; somehow that's in the hands of God. And we're told that each of our lives will end in death, which is evil and a universal punishment, and the world's going to end in the Apocalypse. So, according to one school of Christian interpretion, everything on earth is going to work out terribly in the end, anyway.

"So we shouldn't be caring and pursuing remedies for suffering for the sake of results. We must be doing that simply because it's what we're called upon to do. I think that's really a great strength. I don't think the pro-life movement, for example, could possibly have gone on for fifteen years if it were animated by hope for results. It was simply what its supporters had to do.

"There's also a sense in which I think that in terms of caring for the self, the pursuit of results is depersonalizing. The only real self that any suffering person has is the self in front of him or her right now in the mirror. The idea we have of that person existing at some point in the future without suffering or without retardation or whatever is just an idea, a concept in our mind. So unless we can accept suffering and hold back from this pursuit of results, we lose sight of the personhood of those around us who we claim to care about."

At this point Stephen Post said to Hauerwas, "You've helped us in the Christian community deal with the suffering of a patient with AIDS. But I don't think the kind of argument you pose helps the actual sufferer make much progress in understanding the meaning of his or her condition. Even as a Christian, I don't think the problems of active euthanasia and all sorts of other things can be easily rectified on the basis of these kinds of religious claims."

As the session drew to a close, Hauerwas offered some concluding remarks: "It sounds wise and good to say that to be human is to suffer. That's a start, but we can't end there. I was trying to go on from that point to attempts to alleviate various kinds of suffering and to care for those who suffer. And why is the modern medical project so committed to the assumption that suffering can be eliminated? If I'm right, medicine is the other side of the theoretical theodicies.

"I'm not against the kind of metaphysical issues that Tom Oden raises. They've always been part of the Christian tradition. I'm against the assumption that the church has got to come up with a logical explanation for and response to the sheer existence and power and goodness of God in a way that is separated from its specific tradition."

"It's one thing for Job, in his condition, to ask why God allows suffering." But his question can't be asked from the standpoint of a generalized position of "God cares about us," Hauerwas went on to say. "Salvation is a very specific engrafting into the people of Israel that makes God's promise present to the Gentiles in a way that it would not otherwise be. That salvation really is a social alternative to the world's way of thinking."

With regard to cancer, he said, "the problem is how it functions within a broader framework of the sufferer's life, and how that broader framework functions within a narrative that he or she may not own but should place himself or herself within. That's what we do with children—we place them within a narrative. At one time Christians took their convictions so seriously that they believed they should take their children down to death with them rather than have them raised as pagans. Only against that background can we believe that even children have a place within God's providential care such that their deaths are not without witness and service. They don't necessarily own that themselves. We don't celebrate their deaths, but that perspective gives us a way to go on in the face of their deaths. And it gives medicine a telos to serve other than simply stopping our children from suffering or dying."

GUARANTEEING THE GOOD LIFE THROUGH LAW: THE EMERGING "RIGHT" TO A GOOD LIFE

Controversial attempts to guarantee a good life through technological interventions are often resolved in courts of law. Accordingly, the law became the context of discussion in the next-to-last session of the two-conference series, in terms of a wide range of real-life hard cases: from abortion, to surrogate parenthood, to organ transplants and the medical use of fetal parts, to termination of medical treatment for those in a comatose state, to euthanasia. On what basis are such decisions made—and on what basis should they be made—in courts of law?

In his opening remarks, Neuhaus related the question to the discussion of suffering in the previous session: "One of the things that struck me in our discussion yesterday afternoon—and this is something I hope we'll explore today—is how in this group the alternatives to what we generally saw as the culturally dominant attitudes toward suffering, as a problem to be solved, were almost uniformly and unapologetically very religious/biblical in character. That may be where we are. This may be the alternative. But at some point—and nobody picked up on this—Brigitte said, 'Of course, the language for that alternative is not available to 90 percent of the American people.'

"I think that's wrong. I think it's the natural language of 90 percent of the American people. But they're being educated out of it—by the new class, if you will. And not only educated out of it, but their language and understandings about it are being made increasingly irrelevant by decisions made by others, who do an end run around democratic moral deliberation. And that, I think, brings us very directly to the role of the judiciary, and Bob Destro's paper."

Destro began the day's discussion with comments about developments that had occurred since he completed his paper. Noting that "nowadays landmark cases are heralded by the newspapers as 'landmark' before they're even filed" and that most of the withdrawal-of-life-support cases to date have involved individuals who at least can be called vegetative, Destro described one crucial "landmark case."

"This one involves a soldier who was in a car accident in North Carolina. He's not brain dead; he's not comatose; he's not vegetative. He apparently can respond to questions by tapping people's wrists. But his parents want the hospital to remove food and fluids and let him go.

"This is the next case you're going to read about in the newspapers, and I can guarantee that the standard phalanx of experts and bio-ethicists will be involved," Destro said. He noted that lawyers and judges are not trained in the humanities or in morality, and are dependent on the testimony of experts. Most lawyers and judges, he went on to say, want to "do the right thing" in terms of dealing with the facts under statutory law. But in the more controversial areas such as religion and bio-ethics, they're out of their depth. Here the new-class attitudes of the experts come into play. "Even though I may not be able to define the new class and know there may be some question about whether or not it really exists, I certainly have run into it," said Destro. "And it wasn't until I read Brigitte's paper that I realized that, like Pogo, I might be part of it too."

"Bio-ethics cases," he continued, "are inherently related to cases involving religion and society, whether or not they appear to be on the surface. But there's a very strong notion that religion ought not to be a dominant factor in deciding legal policy. You run into the feeling that somehow attitudes about the value of life and of people are illegitimate if they are religiously based. The general notion in American church-state law is that law and religion ought to be separate, even though the cases recognize that the wall of separation between church and state is hazy and indistinct, varying from case to case."

Starting assumptions, he went on, play a big role in the judicial determination of these cases. "Where does the discussion of bio-ethics and the law begin?" he asked. He discussed two concepts of rights—whether they are to be protected as a duty imposed by society, or whether they are derived from the autonomous individual. He noted that duty imposed from without has been resisted, especially if duty is based on religious grounds. "If the implicit assumption is that human beings should be defined functionally, and the goal is justified by rights based on individual autonomy rather than on duty, then the end point is predictable from one case to the next." The implicit assumptions, the first principles, determine the outcome.

Destro concluded his remarks with a reference to the

"Baby M" case, in which, he said, the court recognized that all the arguments were based on individual autonomy, but that individual autonomy left out the baby herself. "They said, 'This is a baby-selling case, and that's terrible.'" But they added that the legislature could decide that such baby-selling was legitimate on grounds of utility.

In opening the general discussion, Neuhaus saw an element of ambiguous hopefulness in this final point. "It was an acknowledgment by the court that some of these questions are inescapably moral judgments, and they should be made through the democratic process rather than the courts."

Quality of Life vs. Sanctity of Life

Neuhaus went on to note a thread throughout the Destro paper— a discussion of quality of life versus sanctity of life. Myriam Puig suggested "dignity of life" as another alternative. But law, according to Destro's paper, is moving inexorably toward a position in which quality of life takes precedence over sanctity or dignity of life.

"We on the pro-life side don't necessarily need to surrender the 'quality of life' rhetoric to the other side," Richard Stith said. "If there is care for the dependent and the disabled, there is a higher quality of life for all of us in a very staightforward sense. Furthermore, one of the main, traditional, natural-law arguments for human dignity is based on an Aristotelian-Thomistic kind of argument that centers on an idea of quality. Because we humans are rational creatures, designed to love and to reason, we are more valuable than other species. We have a 'quality nature,' so to speak. Quality is latent in our being, in our physical existence, even when we aren't manifesting it. That's the issue, at least with abortion. The most difficult cases are those involving the severely disabled. Even though they can't manifest that quality, they still retain that nature."

"We don't want to be put in the untenable position of saying that we think quality of life doesn't matter," Stith continued. "If I said to my daughter, 'What do you want to be when you grow up?' and she said, 'I want to be just like Karen Ann Quinlan,' I would think that was weird. But she could say, 'Well, you believe in human equality, don't you? She's as good as anyone else.' I want her to pursue quality of life, and yet I want to say that the dignity that we are all pursuing resides not just in who we are or what we do but in the fact that we are human beings."

Noel Reynolds directed attention to the sanctity of life as a possible position behind which a political movement might unite. "I don't believe it's likely to work as a legal concept," he said. "It seems to me that Destro, quite rightly, throws these moral issues back to the legislatures. The law reflects the beliefs of the people. But there is a pessimistic conclusion to be drawn from this. The people tend to be utilitarian in the sense that they see necessity as a justification and are always looking to balance interests. This runs right up against sanctity-of-life considerations."

Comparing cannibalism to the medical use of fetal parts, Reynolds pointed to the gentle way in which cannibalism, driven by necessity, has been treated in the law. When asked how this related to the cases raised by Destro's paper, Reynolds responded that the key question is how such things are justified. "They're justified by arguing from necessity or from the balance of interests. The idea of the sanctity of life has no firm defenders. So the only hope for obtaining the kinds of objectives that we have around this table is to change people's minds. The key thing is the need for a healthy and effective religious basis for society. If we don't have that, I don't think there's any other way to establish these positions."

Jonathan Imber saw the cannibalism analogy as valid, and noted the political difficulties facing the sanctity-of-life position. With regard to the use of fetal parts, he said, "the worm was already in the wood" with organ transplantation based on brain death. "The fact that organ transplantation has become an acceptable innovation in medicine shows that we are already building upon a foundation where the notion of the sanctity of life is entirely jeopardized," he said.

Stanley Harakas disagreed. He saw a radical difference between organ transplantation and the "harvesting" and use of fetal parts. He felt that the one does not lead to the other.

Stith pointed out that cannibalism had been excused in the cases which Reynolds cited but that it had not been "approved" by the law. He compared this with abortion, saying that women who have abortions shouldn't be legally punished but that this doesn't constitute legal approval.

Neuhaus noted, however, that "approve" language is being used today. "Doesn't our society have the capacity to affirm, as a general proposition, that we ought not to kill innocent human beings?" he asked. "If the answer to that is no, we're all in much deeper trouble than we think."

"In the face of that question, everybody is going to say yes," Reynolds answered. "But the question is, How much clout does that have in the law?"

Here James Burtchaell spoke to the language issue: "You're looking for rhetoric that will be more hopeful in resisting what appears to be a new form of legalized savagery. The trouble with 'quality of life' is that it treats the impaired person not merely in terms of function but in terms of *desirable* function—function desired by those at whose cost the impaired person continues to exist. The phrase is used deceptively and perversely when the implication is that the intrinsic quality of the impaired person's life is really being considered. 'Sanctity of life' might as easily be replaced by 'dignity of life' or 'inviolability of life.' But what's at stake is that the decision about that person's existence must not be made with respect to anybody else's interests."

"There are other kinds of language that we have trouble with," Burtchaell continued. " 'Best interest'—that's often decided in a very superficial way. Sometimes it's in my best interest to suffer, but that's not the way it's usually meant. And the language probably classically used by Pius XII, the language about 'ordinary' and 'extraordinary' means to resolve this problem, generally has been reduced to a question of wherewithal. 'We just can't afford dialysis for everybody,' it is said. I think the issue is deeper than finding the right word."

Potentiality as a Criterion for Preserving Life

Neuhaus invited the group to focus its attention on the fact that "in general people who favor abortion on demand are also disposed to favor euthanasia and the termination of life that's allegedly not worth living. But they use an argument in the second case that they're not prepared to use in the case of abortion: that it's the performance rather than the being of this person that matters, that in a permanent vegetative state the potential for performance isn't there. That's one reason why such a life isn't worth living. But if that's the criterion, consistency requires recognizing that the fetus *does* have that potential. It would be interesting to explore whether within the pro-choice, quality-of-life side of this argument that is as blatant a contradiction as it seems."

"Potential appears to count," Richard Stith answered, "but only after it's been manifested. Once a person has talked and loved, what matters is whether he or she can recover those capacities after serious illness or injury. But the unborn hasn't yet been able to talk or love, so that potential doesn't count. It seems to me an ad hoc argument that doesn't have any intuitive force."

"I think it's not as blatant a contradiction as that," Gilbert Meilaender countered. "You'll find in the abortion instance that the argument flows back and forth between quality-of-life considerations and the point that the life in question is peculiarly dependent on the mother. That's the claim that will be made at the point where you try to press the issue that there's potential here that there isn't, say, in someone in Karen Quinlan's circumstances."

Robert Destro added, "There's also one other factor regarding this potential, as Jean Elshtain pointed out in last month's conference. They talk about the woman's lost potential in the abortion case, too."

"So you have rights in conflict," Neuhaus said. "I don't know how much we want to explore the 'potentiality' and 'personhood' questions, but it is perhaps important to note that the questions that arise at the entrance gates—the abortion questions—are quite different from the ones that arise at the exit gates—the euthanasia questions—in that in euthanasia debates (and also in the court cases dealing with euthanasia) the potentiality issue is clearly and explicitly raised and pressed. And maybe all we're noting is that there seems to be an incoherence here, because the same kind of argument isn't expressed in abortion debates and cases."

Personhood as a Criterion

Stanley Harakas raised a question about using the term "personhood" to imply nothing more than biological existence. "In many traditions the idea of personhood is somewhat plastic in the sense of growth. It is connected with a teleological idea of personhood as something that constantly grows and needs others for its fulfillment. Personhood implies community." He suggested that in the present context the designation "human being" might be better than "person."

"I don't think it's appropriate to talk about the fetus being a person," Stanley Hauerwas said. "It's more determinative philosophically to say, 'This is a child.' 'Child' is a more primitive moral notion than 'person.' As soon as I let someone get me to use the language that says this fetus is an individual, and I have to come up with defining characteristics in order to discern if it is a person, I've lost the game."

"Child is a relational term, a more accurate term," Neuhaus added.

Jean Elshtain recalled the case of a doctor in Boston, a case in which lawyers lamented the fact that ordinary people in the jury might think in terms of "unborn child" rather than in medical language such as "product of conception." In ordinary discourse people do think this way, she noted.

"So does the physician," added Bernard Nathanson of Bernadell, Inc.

"The folks we're talking about," Elshtain continued, "live within this liberal society. They have access to this language. They obviously go beyond the abstract language of personhood, and they have resources to draw upon to sustain this alternative framework. Can't the law incorporate those received understandings out of this alternative tradition? Must it always move into the abstract language?"

"I think you're right, Jean," said Neuhaus. "Just as we were saying yesterday, there are languages within the society for alternative understandings of suffering, of personhood, and so on. But in this specific case, isn't Bob Destro telling us that, given the inexorable, rigid character of *stare decisis* and one thing leading to another, the legal system has locked itself into a narrow linguistic track that isn't in conversation with—indeed, is relentlessly opposed to—that alternative set of language and vocabulary?"

"That's the question," responded Elshtain. "Is the law that closed?"

Equality of Right

Thomas Oden suggested equality of right as a criterion, "the right of the defenseless and vulnerable as well as the able to protection of life. That's a language I think the courts can understand."

Hauerwas saw problems with egalitarianism. "I'm against

egalitarianism because it's correlative to liberal presumptions about individuals in a way that's destructive of important discriminations. Egalitarianism is going to destroy you, because what you're trying to equalize is an abstraction."

Robert Destro pointed out that the legal principle of equal protection is different from egalitarianism. "But," he went on, "the courts subvert even that, because, as Stan pointed out, everybody isn't situated equally. One's duty to some people might be different from one's duty to other people. What's more, some people have translated equal protection into some kind of neutrality, and the law is struggling with that, too. The equal-protection principle is subverted all the time in areas in which there's a desired social outcome. In the area of women's rights, for example, the reigning idea seems to be, 'There aren't enough women doing x or y, so let's violate the principle and discriminate against men for the time being.'"

"From the standpoint of the theory of law," Noel Reynolds said, "what we're dealing with is a confusion between formal equality and equality of result. The whole idea of law requires a formal equality—that is, that everyone be treated equally. But what we get is a perversion of that—the involvement of the government in distribution—and all of a sudden it means equality of result. Even the idea of equal protection was originally a notion of formal equality, but in affirmative action and other such programs it always turns out to be equality of result, which is a perversion. The two ideas aren't consistent with each other. But you can't be against formal equality before the law."

Hauerwas disagreed. "That's an appeal to fairness. It's not clear to me that an account of fairness requires an account of equality."

Neuhaus reminded the group that the equality issue arose in terms of the distinctions that can be made relative to the handicapped, those who require a different understanding of protection than is ordinarily meant when we talk about equal protection under the law.

"And there's no question," Destro added, "that the notion of individual autonomy is in conflict with that growing notion of duty to protect those who are vulnerable, because to the extent that a duty is imposed on an individual, he or she is not autonomous anymore."

Duties to the Disabled and the "Threshold" Issue

Hillel Fradkin pointed to a difference under the law. There is one category of disabled people—those in wheelchairs—for whom the law recognizes an obligation to improve the quality of life. But for another category of disabled people—the brain-damaged—in another set of cases, one of the proposed remedies is to kill those persons to improve their "quality of life."

This, Neuhaus suggested, is moving from a functional definition of disability—what a person can do—to an ontological definition—what a person is. "When persons are handicapped functionally, their entitlements increase under the current development of law. We owe them additional duties. But when their dysfunction reaches a certain point, we say a threshold has been crossed, and the functional definition moves to the ontological definition, and we say that these individuals are no longer persons. That is, in fact, the direction in which the law is moving. At some point, then, isn't the law going to have to answer the big metaphysical-ontological question: What is that 'X factor' that marks the threshold which, when crossed, moves a person from being disabled and therefore entitled to increased protection, to not being a person?" In the general discussion that followed it was suggested that the threshold may be reached when a person can't perceive that he or she is disabled, or when nothing can be done about it anymore. Richard Stith saw it as the point at which death is perceived to be "in the best interest" of the person.

"That threshold is a measure of legal duty," added Destro, and Neuhaus reiterated the bind in which the courts are caught.

Gilbert Meilaender disagreed. "If the criterion is something like the capacity for human relationships, why is it a bind?" he asked. "I think it's a mistaken criterion, but talking about enhancing the capacity for human relationships really is different from talking about having no capacity for human relationships. This criterion is wrong, but I don't think it's unreasonable."

"You're saying that in fact the courts have given a kind of answer, and although you don't like it, they can give any answer they want and justify it?" asked Neuhaus.

"I think that answer resonates with the way a reasonably large number of people feel. Just 'any old answer' wouldn't do. I think this one is persuasive to many people and therefore is politically persuasive," Meilaender replied.

Brigitte Berger asked why an individual who rationally recognizes that she wouldn't want to live in a vegetative state for the next twenty years can't make such a decision—instead of having the courts make it—and guard herself against having others take extraordinary measures to sustain her life. Bob Destro suggested that this is an individual-autonomy argument that could as readily be applied to a stock-market decline. "If you don't like the conditions, you have control over your life."

Individual Autonomy and Community

At this point Bernard Nathanson brought up the recent Dixon decision of the Canadian Supreme Court as an illustration of the march toward absolute individual autonomy. "It's based on the Canadian Charter—this new constitution—which guarantees 'security of the person.' Those four words, applied to the mother, are used to strike down any existing restriction on abortion in Canada. There's no balance of rights at all. In the entire decision there's only one statement recognizing that there may be some 'interest' on the part of the fetus, but this is clearly outweighed by the 'security of the person' guaranteed in the charter. I think now the Canadians are clinging by their fingernails to the edges of the 'slippery slope.'"

But Stith suggested that the Canadian case might not be as definitive as Nathanson feared.

Christopher Lasch noted the tendency to define selfhood in disembodied, abstract terms, with Artificial Intelligence as the logical culmination of the ideology of individualism. "The notion of embodiment," he said, "can also refer to embodiment in a rich texture of social relations. There embodiment has a particular history. Selves can't be detached from these relations without losing the very essence of selfhood. Hauerwas's distinction between the child and the person is a nice illustration of that point, when you think about the radically different implications of those two ways of talking about an embryo. What are the social roots of the abstract conception of selfhood that would make some wish to talk about 'persons' rather than children, fathers, mothers? Here one would want to consider the whole history of liberalism, the tradition to which we owe the concept of the disembodied individual.

"And one would want to talk about the many ways in which liberalism is indissolubly associated with capitalism and the modern market economy. So the conception of the abstract, disembodied, free-floating, freely choosing, contracting, decision-making, rational, autonomous individual is a description of personhood in a social order in which persons are related to one another only through the market, as buyers and sellers. That this is the only relation which the state and the law can take account of is one of the key assumptions of the liberal tradition. This is why liberalism has always found it so hard to talk about the family, about relations that are not contractual—though we see the contractual tendency here, too, in the notion that children have contractual rights. This is the kind of thinking that underlies the separation of church and state, another foundation of the liberal order, and the exclusion of moral issues from public discourse on the grounds that they are simply questions of individual preference that can't be allowed to influence public debate."

Individualism and the Free Market

"I don't see how you can criticize this particular conception of selfhood without also criticizing the market model of society and social relations," Lasch continued. Pointing to the relationship between cultural conservatism and economic liberalism (the configuration that elected Reagan to the presidency), he professed to be puzzled by that fact that the people most concerned about the issues being discussed at the present conference were often firmly committed to the idea of the free market.

"My objection to the new-class theory, which is itself ideological," Lasch said, "is that it seems to paper over the fact that all the things that we are objecting to are rooted in a liberal tradition that, to put it crudely, is an ideological reflection of free-market capitalism. The new class, which is alleged to have developed in opposition to bourgeois culture, may have looked rather bourgeois for a long time because the full implications of modern individualism were muted by the persistence of older cultural, moral, and religious traditions. It's only as those traditions have become increasingly attenuated in the twentieth century that the underlying implications of liberal individualism, as we see them in the quality-of-life ideology, have been fully revealed for the first time, unqualified by the residual remnants of earlier traditions. However, I think it's much more useful to see this ideology not as the characteristic self-interested expression of a new social formation but simply as an outgrowth of bourgeois culture."

Berger resisted this line of reasoning: "I don't think the new class is directly an outgrowth of the market, organized under capitalism. The Soviet Union, which isn't organized under the market, has a new class as well. This is part of the modernity syndrome, which also produced individualism, which also produced the market. Your analogy is very wrong."

But Neuhaus found highly relevant to the subject at hand the claim that "these legal, social, cultural, and medical directions, these attitudes and the practices that come out of them—from abortion to euthanasia to the farming and harvesting of fetal parts—are in some way a product of a liberal individualism that is necessarily correlated with free-market capitalism."

Jean Elshtain added what she called a "quick commercial footnote" to what Lasch was talking about. "It strikes me that one of the reasons we in this society moved so far so fast in the direction of buying and selling fetal parts, genetic engineering, and all the rest is that there's money to be made in these areas. The number of bio-genetic engineering companies is staggering. It's that runaway commercialization, and the fact that we can't figure out how to put any brakes on it because it's the nature of our society, that helps explain why we're at the impasse that we are right at the moment."

"And don't forget the universities," Edward Gaffney added. "They make big bucks, too."

Gilbert Meilaender addressed the market issue. "I don't see why we need to think in terms of either all market or all non-market," he said. "There may be certain kinds of things that should be dealt with by market forces, and other kinds of things that shouldn't be. We have to look at each one, sort out the argument, and see.

"The other thing I want to say is that I don't much like personal language and I don't much like autonomy language, but I don't mind individuality language—'individuality' is a good Christian kind of word. And the notion of embodiment in history, although I think it's fruitful and I like it in a lot of ways, also has certain dangers. If we have a relational concept of selfhood like that, and we have a society like ours that has increasingly lost any sense of the transcendent, then the largest totality within which relations might take place is the social one, and I think that's very dangerous. A concept of individuality, which I think is always finally over against the great Individualizer (and I admit I don't know how to work this into public discourse), is very important to retain in order to guard against social totalitarianism."

Individualism and the Patient

At this juncture Neuhaus summarized the discussion thus far: "Bob Destro started us out on the question of the quality of life versus the sanctity of life—or 'dignity of life,' as Myriam suggested. That was the agenda that was set for our morning session. Our discussion has moved—perhaps with logic—to the question of individualism and community, and then to Kit Lasch's particular connecting of that question to the question of capitalism and contractual or exchange relationships. Are Bob Destro's original question—quality of life versus dignity of life—and his dismal conclusion—that in law the quality of life is inexorably proceeding to a dominant and unchallengeable position over dignity of life—necessarily related to the question of individualism versus community?"

There were several yeses from the group.

Destro said, "Kit raised a point that I think is really central to all of this. He talked about the relationship of the self to the whole and the division of the self from the organic whole. If we relate this to what's going on, we realize that radical individualism is gobbling up the person who is the individual." Destro went on to say that people must be taught that the kind of autonomy they have been seeking—divorced, as Lasch has pointed out, from the older cultural, religious, and moral traditions—will result in their demise. "They must be told, 'Through your radical individualism, as you atomize from the whole, you will lose any connection, any duty. All those rights you have are really dependent on others' carrying out their obligations to you. If those obligations aren't contractual, you just pray that they'll fulfill them for you, and as soon as they decide they're not going to anymore, you're gone. If you're useful, they'll use you. The co-matose patient is going to die anyway, so they'll use him.'" Press reports about a French doctor asserting the right to experiment on such patients were cited.

Discussion followed about whether the loss of individu-alism is really at the root of the helplessness of the comatose patient, or whether, as Meilaender suggested, the individualism of the patient could be asserted as a protection.

Neuhaus asked, "If that patient is—let's not even call it something so elevated as capable of relationship—simply cap-able of perceiving and asserting, of saying back to a doctor, 'You can't do that; I'm an individual,' then clearly that person's status as an individual would be under protection."

"No," Lasch responded. "The whole point of Bob's paper is that that's precisely what isn't protected."

Meilaender sought to clarify his point: "What's happening in these cases is complicated. Having decided, on essentially so-cial-relational grounds, that what we've got here is a life not worth preserving any longer, we have to find a way in law to accomplish that. And we find a way in law, through substitute judgment, to assert that individual's own autonomy: 'He would have chosen death if he could.' But that's the *second* move that comes after the other decision—that this is not a life worth preserving—which isn't grounded in liberal individualism. It's grounded in an un-derstanding of human personhood as determined by a capacity for relationships within human society."

Lasch continued to disagree. "I just don't see that empirical-ly. Of course individuality is an important value. It's part of what I was talking about, part of what one talks about when discuss-ing the sanctity of life. But, as Bob said, that itself is the final casualty of this process. Individualism is self-defeating."

Stanley Hauerwas noted that the term "individual" is of rela-tively recent (seventeenth-century) origin, but Neuhaus pointed out that behind it, in terms of biblical grounding, "is the prophet-ic insight that each will be held accountable for his or her own acts before God, that each life stands over against the community. This notion is deeply grounded. And the task ahead of us is to preserve what are—I am prepared to say—the providentially provided fruits of democratic and pluralistic governance, which are grounded in a liberal-individualistic tradition to a large extent. We should do that in a way that prevents the tradition from consuming itself. It's consuming itself right now; one indication of that is what's happening to people at the edge, to those who are most vulnerable, who are the primary focus of our conference."

The Constitution and Providing for the Needs of Others

James Burtchaell noted the reference in Destro's paper to the law's dealing with "active and passive steps . . . taken with the specific intent of ending one's own or another's life when that life has lost meaning, quality, or value to oneself or others." He commented on the way in which "lethal behavior which would never have been permitted in an active mode is now being entertained in a passive mode. That's raising more clearly issues that haven't been noted for a long time: the degree to which and the reasons whereby one may, with legal impunity, ignore another's need."

"The duty to provide for other people's needs is nowhere in our constitutional tradition, nor is it in the Enlightenment philosophy that lies behind our constitutional tradition," he continued. "To free people from the ravages of unjust rulers, to give them appropriate liberty, the constitutional tradition, which grew out of the various revolutions, generally alienated 'the citizen,' later called 'the person'; that isolation from neighbor was the cost of such liberty. Libertarian protection for the individual carried a much higher cost than was thought, because nothing in this constitutional tradition spoke of a person's inherent duty to provide for someone else's need.

"The way it was handled was, of course, through contract, but also by statute. Statute could create an obligation to provide. Kinship was thought to do so naturally, but subsequent events have raised many questions about that. The obligation to provide hasn't been so much of a problem until recently. Now for the first time the Supreme Court has been so rigorously in the hands of enough articulate legal positivists that the attempt has been made to make the judiciary run on the Constitution alone. The Constitution was always only a part of our way of life; it was surrounded by other shared understandings, including the understanding that we had a duty to provide. But when that's removed, the Constitution becomes a savage document which allows me to stand by and watch somebody bleed to death. Now, when people become a burden to me, I can simply fold my arms and watch them get out of the way."

"Or, even more actively, you can terminate them," added Neuhaus. "This is a valuable point," he continued. "Originally the Constitution was seen as a rather limited thing, related to the notion of limited government, surrounded by all kinds of other communities and other understandings of obligation and relationship. But when the Constitution is construed as being *the text*, so to speak, by which we live our lives, it's a very brutal one, because it's based on a set of essentially self-serving, contractual-exchange relationships. I think what we can agree on—despite our different attitudes toward market-economy capitalism—is the unhappy fact that the minimal conditions of liberal democratic governance have become for many people today the maximal conditions. What we would disagree about is *why* that happened."

"I'd prefer to put it slightly differently," Burtchaell said. "The Constitution is a polemical and corrective document. It has strong emphases intended to remedy terrible human mistreatment. It's not any more than that. But now it's being treated as a judicial statement about how we are to live our lives together. And there's nothing in the Constitution that would make anyone treat anyone else as a neighbor."

Meilaender dissented. "I'm not persuaded by Jim Burtchaell that the defect lies in the legal system. In that quotation from Bob Destro's paper, the central ethical question is about 'active and passive steps . . . taken with the specific intent of ending one's own or another's life.' There's never been legal permission to *intend* to kill someone. What's going on here has very little to do with those larger questions about what the Constitution was originally intended to do. What we are trying to figure out is whether there are passive ways to end a person's life that really still amount to intending to kill that person. Intentional killing has never been legally permitted. It's just that the courts are no longer interpreting passive euthanasia as intentional killing. That's the issue."

"Withdrawing of food and fluids is pretty explicitly intended to terminate," Neuhaus commented.

"But those who withdraw food and fluids deny that that's what they're doing," responded Destro.

"It's readily admitted in many cases that the withdrawal of

sustenance is the first form of direct intervention to terminate," Neuhaus said. "That's why the tubes are taken out."

"I don't think any court has approved the withdrawal of food and fluids on those grounds," Stanley Hauerwas objected.

"But the courts are close to doing that," said Destro. "And that's why I think Jim is right when he says it's dangerous. The common-law decisions are important, but they're mixed with constitutional analysis. The courts are always going to use the Constitution as a trump card."

Noel Reynolds objected to Burtchaell's interpretation of the Constitution. "The answer to the kinds of things you want to see in the Constitution is that the document structured the ways in which the Congress, in concert with other branches of government, could impose certain duties on the population. And it's certainly possible that all the things you want to talk about could come out of that."

"Congress could impose duties, but with very sharp limitations, which the Bill of Rights most dramatically illustrates," put in Neuhaus.

"Congress could require people to come to the aid of a dying person," Reynolds said.

"But there's nothing in the Constitution to make us think that even without such a law we already have such an obligation," Burtchaell maintained.

Bernard Nathanson asked Burtchaell, "If you rule out appeal to the Constitution and appeal to the courts, where do two parties who have a substantial moral disagreement go?"

"The common law always assumes a shared understanding of how we have to live in community," Burtchaell answered. "We can appeal to the legislature and also to the judiciary and argue any question of justice or equity. We can go to the House of Representatives and argue any question of justice or equity using religious grounds, the claim of necromancy, or our own private sense of right, and the representatives will dutifully listen. But legal positivism says we can only argue what's on paper somewhere. Legal positivism as applied to constitutional decisions— we're really talking about the Supreme Court—is encouraging the courts to deny themselves the right to any of the shared convictions about how we are to live in society."

At this juncture Kenneth Myers joined in. "You wouldn't say

that *Roe v. Wade* was based on the articulation of shared convictions? That the Supreme Court was saying that this is what the shared values of the society are? It seems to me that the radical judicial positivism of the past decade has been a reaction to reading things into the Constitution based on an assumption of what values were shared." Discussion followed about the extent to which the decision may have been based on shared values or an interpretation of the Fourteenth Amendment.

Destro concluded the session by saying, "That may be a good place to end. Blackmun thought he was reflecting the sense of the Fourteenth Amendment. The theory on which he was relying misses the natural-law/common-law tradition and centers on legal positivism. If we go back to the notion that the Constitution is a positive document, that it covers what it covers and it doesn't cover anything it doesn't cover, then all those cases that we call substantive due-process/fundamental-rights cases go right out the window as not based on the Constitution in the first place. Those cases are left to more of the common-law tradition in the state, where they rely on both the legislative and the judicial ways of looking at the shared values of the community."

Destro spoke finally of the courts' attitude toward the role of religion in society. "We have to go at the foundation," he said. "We have to go after each one of these very fundamental propositions. The degree to which people start to yell and scream and call us racists or Neanderthals is the degree to which we've got to keep going."

GOD, THE BODY, AND THE GOOD LIFE

The individual autonomy of the self, promoted by the new class and embedded by the courts in a seemingly inexorable progression of case law, had emerged in the semi-final session of the conference as a key element in undergirding the new eugenics. Paul Vitz began the final session by summarizing his paper, which, he said, calls for serious reconsideration of the autonomous self, independent of body, independent of others, and independent of God. Much of the pathology of the late modern period comes from this autonomy, he contended. His paper re-argues the case for *connection* in these three areas (body, others, and God), particularly in the first and the last, the middle area being not unimportant but addressed elsewhere. "In my paper I argue that we're not independent of our bodies, and we're not independent of God. In particular, I seek to sketch out a way in which a link to God can be intelligibly presented and dealt with today," Vitz said.

For the first time in many years the intellectual community is potentially open to the question of God's existence, Vitz claimed at the close of his summary. "If we're ever going to recover from modernism and its pathologies, which we've been talking about so eloquently and sometimes depressingly during this conference, an important item on the agenda is to make the question of God an important issue for the intellectual community. In many respects atheism began the modern period, and it seems to me that to move out of atheism is one of the ways to end it. I think the time for that is propitious."

Before initiating discussion on Vitz's paper, Neuhaus noted that the question on which Vitz had been asked to write was, What does it mean to say that a society is suffering a disease of the soul? "When we look at all of the things we've been talking about in terms of the return of eugenics," Neuhaus went on, "we realize there's something wrong here, there's a disease of the soul. And while Paul never uses that term in his paper, I think he addresses the issue admirably. It's not simply the God question but also the man question, which can't be answered except in relation to the fullness of embodied man and God."

God and "the Transcendent"

Jean Elshtain opened the general discussion by asking Vitz a question about the language of the transcendent that he used in his paper. "Are 'God' and 'a transcendent spiritual world' the same thing? Don't you lose some specificity when you talk about the transcendent? And don't you think that in our culture 'the transcendent' is up for grabs? We have the 'harmonic convergence' and crystals and all those books by women with one name. That language of 'the transcendent' puts you smack-dab in the middle of the 'New Age.'

"The other example of that is the language of 'religious experience.' You didn't say 'religious life' or 'religious faith' or 'religiously formed character'; you said 'religious experience.' It seems to me that that language trivializes. I know that's not what you intend, but I'm wondering about that use of language."

Vitz said he was aware of the problem but was purposely keeping the argument very broad.

Neuhaus asked Elshtain, "Are you open to the possibility that—as in the early era of the Christian church, the first couple of centuries, when there were Shirley MacLaines running amok all over the place—we might be entering a similar period, and Vitz's language might be the price to be paid for the desired collapse of what we're running from?"

"I think that's a very optimistic reading," Elshtain replied to Neuhaus. "We're talking about disease of the soul while you're smuggling in more of the illness."

"I think it's critical to keep in mind the audience this paper is directed to," Stanley Harakas put in. "You're trying to open up a potential realm that's been closed, Paul, and I think it's legitimate."

"That was my strategy," Vitz confirmed. "First I want to get my audience to accept the transcendence, and then we'll fight over the nature of that transcendent realm. But until my audience is willing to admit to the existence of the transcendent realm, I'm not going to debate the other issues."

Religious Experience

Brigitte Berger agreed: "It's legitimate to begin the religious quest with a definition of the human relationship to the supernatural. The biblical tradition works up-down, but today it's very important to work from down-up. Getting at this is something that has to be done to some degree through experience."

Vitz added, "Religious experience is the fundamentally American aspect of religion going back to Jonathan Edwards. I chose that aspect deliberately because I thought that's where I could meet an American audience."

Stanley Hauerwas challenged the approach. Referring to his own background as a Methodist who had had "enough religious experience by the age of fourteen to last a lifetime," he said, "We also have an intellectual history on this. Once Descartes and Newton excluded God from the realm of nature, the way theologians then tried to show the continued relevance of theological lan-

guage was to show that it wasn't possible to avoid giving an account of God as correlative to human subjectivity. God isn't present in nature or history, but at least God is still present in human
experience. That subjectivizes God, leaving us open to the charge
that our talk about God is a projection of ourselves. So many of
us are deeply suspicious of beginning with the language of experience. It doesn't give adequate purchase on the character of the
God we worship, who moves the sun and the stars. If we begin
theology in the language of experience, ultimately—as Barth
taught us—we end up talking about humankind in a very loud
voice."

Neuhaus pushed him. "You wouldn't deny—or would
you?—that Paul is correct in challenging people to consult their
own experience—to 'press the envelope' of their own experience, so to speak—in order to discover that there are indeed encounters with experiences ('experiences' is an inescapable word)
for which the best language is the language of the supernatural,
the transcendent, God?"

Hauerwas persisted in resisting the assumption that "if you
rub your experience hard enough, you'll discover God."

"Paul doesn't say that," countered Berger.

Neuhaus added, "Even in the Barthian understanding, you
encounter the Word, and in that encounter you say, 'I think that's
true.' And then you check it out, in the sense of your understanding of Thomas's proofs, and you discover that there is
coherence in the construction of reality that the Word proclaims.
Being open to that—isn't that what Paul is talking about in terms
of 'seek and ye shall find?' "

"All I wanted to do was to call your attention to this background history that makes many of us in theology very hesitant
to use the language of experience," Hauerwas concluded.

Gilbert Meilaender observed that there is a sense in which,
historically, the pagan was a person on the way to Christianity
and open to Christianity as God's creature, perhaps in a way the
secularized person of today is not open to it. "To get at the way
in which this person is 'open to it' is important, and there are
certain kinds of experiences that make such a person open. I
wouldn't want to give that up. There's an assumption at work,
behind 'seek and ye shall find,' that God might be active, that
there's somebody there who cares about making contact and is

trying to make contact. There's a rather definitive biblical-Christian notion at work here. It's not just 'transcendence.' "

"That's very closely related to Barth," Neuhaus suggested, "to the whole notion that one can't hear the law unless one intuits something of the gospel behind it." Meilaender agreed.

James Burtchaell voiced some concern over the direction the discussion was taking. "In a group that has so consistently, and on largely Christian grounds, decried the depleted and withered experience of a human life which is not in community with other human beings, I think it would be ominous if the only voice that has explicitly said we also stand to wither if we are not in community with God were simply to be criticized for that. We have not yet said explicitly that there is a great deal to be lost if we are not in an expressed relationship with God."

Hauerwas agreed, saying that he had spent too much time raising questions without making this affirmation. "One of the central theological challenges before us today is the result of our having given the so-called sciences the right to think they can operate adequately without any claim about the reality of God making a difference in how they operate. On that you're exactly right. The creation-science people, despite all the bad ways they've got the issues, are right when they say that if we teach biology as if it makes no difference whether or not the world is seen as created, we are in fact teaching atheism. When we teach psychology as if people are nothing but bundles of chemical interactions, we are in fact teaching atheism. Until Christians are able to reclaim the reality of God as intellectually compelling for sociology, we're in deep trouble."

Body and Mind, Personhood, and the Ending of Life

Meilaender raised a question about the relationship between the body-mind connection that Vitz discussed in his paper and the personhood issues of the morning discussion.

Stanley Harakas commented that our responses to bio-ethical questions are paradoxical. "On every question, when it comes down to a choice, what we opt for as the supreme value is something understood as 'personhood.' On the other hand, we live out our lives as if we were not persons but merely bodies that need satisfaction, titillation, and that sort of thing. I think that the idea that a human being is just 'person'—or 'mind,' as a substitute word—leads to an openness to all the things we've been decrying here in this meeting."

Richard Stith added, "At the very least, Paul's point may lead us away from some of the dualism we were talking about this morning. The question 'Is this bodily life doing the comatose person any good anymore?' shouldn't make any sense in light of Paul's paper, because the body and the person can't be separated. That doesn't resolve all the issues, but it gets us over a lot of hurdles."

"Richard's point is a good one," said Neuhaus. "The logic used in the euthanasia stuff implicitly assumes a kind of immortality—the notion that death might be good for a person. The courts talk about death as 'good' for the person—an interesting metaphysical assumption."

"But doesn't that mean one of two things?" Meilaender asked. "Either the person is suffering intensely, and the shorter the period he or she has to endure, the better; or the person has been reduced to a state inconsistent with human dignity, and the shorter the period of time he or she has to exist in such a state, the better. I don't think that the idea of death as good for the person necessarily presumes continued existence."

Hauerwas said to Meilaender, "Given your position, you don't want to say, 'Where there is body, there is mind.' You want to say, 'This body is mind.' You don't want to make that distinction operative any longer. You don't have mind separate from flesh. So you don't want to say, 'Sometimes there's a living body that isn't one of us.'" Meilaender agreed.

Neuhaus said, "There are living animal bodies that aren't 'one of us.' So there might be a living, genetically human body that isn't 'one of us.' How do you respond to that? Why don't you find that satisfactory?"

Meilaender replied, "I don't find it satisfactory because certain concepts of the person are deeply grounded in Christian understandings—the way Christians worked it out. To say the kinds of things they wanted to say about God and about Christ, they had to say simply that a person is someone who has a history—that is, they didn't start with any set of characteristics like certain kinds of capacities. If we follow that example and start with that history rather than looking to see if the characteristics are there, we simply have to interact with that body. Where we've got a living human body, the presumption is that that body is one of us, and that we should cherish it, care for it."

"And there has to be some compelling reason for saying that it's not?" asked Neuhaus. "It's not enough simply to say that the body is not in conscious relationship or communication?"

"Right," said Meilaender.

"And to the person who says that is enough, you simply say . . . ?" asked Neuhaus.

"I think there are also 'slippery slope' kinds of arguments here," Meilaender said. "We may get into gradations of capacity for human interaction, for example. There are a variety of arguments. But for me, ultimately, the metaphysical question turns on whether the person is identified by certain kinds of characteristics, or the definition of the person starts with the Christian understanding."

"It seems to me," Neuhaus countered, "that according to the biblical understanding of the person, one isn't prepared to say 'This is not one of us' even at the point of death. The dead person continues to be one of us because we believe in the resurrection of the dead. Death is only the point at which the deceased stops *doing* certain things."

"The person no longer exists among us in this temporal life. How about that formulation?" Meilaender suggested.

"The temporal and the eternal," Neuhaus said. "The question we got onto this morning, and that we're back on now, is where the threshold is. The farther we push it, the more clearly we can see that it is inescapably going to require what is identified as a religious answer."

"I'm not sure I'm ready to buy that," Meilaender responded, "and I wondered about that this morning, too. You see, when a person is totally brain dead, I don't have problems with the decision to simply turn off the respirator. My belief is that a person is always in relation to God, wittingly or unwittingly. But I'm not willing to let that paralyze my capacity to make certain determinations that we must make—namely, when what we have is a corpse and no longer a human being. And I'm not sure what this religious move is supposed to do here. If it simply cautions prudence in our judgments—that we have to be very careful about assuming that simply because an individual is no longer capable of interacting, he or she is not a person—that's one thing. But that can easily lead to a sort of paralysis of the ability to decide when an individual is a corpse and not a person any longer."

Hauerwas agreed, saying that the crucial thing from the Christian standpoint is respectful treatment of the corpse.

Organ Transplantation and the Use of Body Parts

"Would you do experiments on that corpse?" Myriam Puig asked Meilaender.

"Would you remove the organs?" Jonathan Imber asked.

"Under certain circumstances. And we'd have to sort out the circumstances. With the prior consent of the donor, I might extract the organs for donation," Meilaender said. "The problem with what's being done with anencephalics now is not that their organs are being extracted for donation but that they're being kept alive for that purpose."

"But most organ transplantation today works on exactly the same assumption," Imber pointed out. "Pronouncing a person brain dead doesn't mean automatically turning off the respirator; if there's any option to use the organs, the respirator is left on."

There was considerable discussion of this issue, with disagreement about whether the anencephalic infant and the brain-dead person are comparable. James Burtchaell posed the case of a dying man who might request that he be kept alive by artificial means so that he could donate one of his kidneys to his twin, who would have to travel from Malta to receive it. Some saw comparability, and others did not. The question about whether or not such a decision could be made by proxy by a wife or by parents of an anencephalic child was also discussed, and again there was no agreement.

Neuhaus began to summarize. "Richard Stith asks what's wrong with using the organs of an anencephalic child. Gil Meilaender assumes that that violates the child's dignity in an illegitimate way. But Bernie Nathanson raises an important issue about keeping alive numbers of bodies in order to use their 'spare parts.' What kind of principles do we think would preclude that? I assume most of us find this prospect abhorrent."

"But, Richard," Imber objected, "look at our current situation. Already on our drivers' licenses we can give permission for our organs to be used in transplantation. We've already instituted that system. What you object to is some science-fiction, visionary picture of this kind of thing ultimately leading to the subversion of a charnel house. But I think the principle is already in place. Now the question is, Are we going to live in a society that encourages it further, or are we going to say, 'Perhaps liver transplants for very young children are in order, but they shouldn't be done for those who have suffered the ravages of alcoholism'?"

"So you're saying that the floodgates are open, and it's just a question of trying to contain the flood," Neuhaus said.

A Warehouse of Body Parts?

"We're not going to stop with liver and pancreas and kidney transplants," asserted Bernard Nathanson. "We're undoubtedly going to go on to gonads for those whose sexual prowess is failing, and to new skin and hair for those who want to remain younger looking. And right on down the list. Nothing will be lost. All of the parts of these bodies hanging on the wire will be used because there's a commercial market for them."

"Are we indulging in a grotesquerie that we don't need to seriously concern ourselves with?" asked Neuhaus. "Or, in fact, as Jonathan Imber is saying, is this where we are, and is it simply a matter of time until the commercial advantage and other dynamics catch up with a principle already established?"

"The question is whether this is an issue of principle or an issue of degree," Burtchaell said. "I think the best perspective from which to enter it is that of the proxy—the protective, intelligent, caring proxy. Let me put myself in my wife's position. Bernard over there is trying his best to get what's left of me for beneficial purposes, and my wife's not so sure she wants to sign it all away. She sees great sense in waiting for the brother from Malta. She knows this will extend whatever my experience is for another couple of days. It seems fitting. But when Bernie wants my wonderful scalp for Jonathan Imber, who thinks his chances for tenure at Harvard would increase if only he had a pompadour, then I would expect my wife to say, 'That's it.' But that's not a decision of principle; that's a decision of degree."

"Exactly," said Neuhaus. "But let's change it a little. Bernie has pushed us a bit farther down the slope to a point where your wife would say no. But let's say that a number of people less conscientious than your wife have already implemented the use of all organs. It's considered a great breakthrough, has become socially legitimated and legally secured—because, as Imber says, the principle has already been established way back there. And, as Bernie says, the commercial factor comes in. The process won't be called 'selling your body parts.' There'll be a term for it— something like 'intergenerational children's endowment.' This euphemism has an altruistic dimension. In this situation, the pressure is on your wife to say yes to the various uses of your body parts. The practice has become socially acceptable, altruistic, and morally approved. And the kicker here is the suggestion that we've already made the decision, and this is the more or less inexorable outcome somewhere farther down the slope. That's why I think we have to get back to Imber's challenge, which," Neuhaus said to Meilaender, "you were trying to grapple with earlier. And I think you caved in much too quickly."

"No, I don't think so," Meilaender replied. "That's why we want to make it a matter of principle and not simply a matter of degree."

"What principle?" asked Neuhaus.

"The principle," Meilaender answered, "is that the treatment decision should be made on the basis of what's appropriate treatment for the person, not on the basis of keeping the person alive for the sake of harvesting organs. All I said was that my wife could make that decision, and I had given her a durable power of attorney."

A Therapeutic Criterion for Medical Treatment?

"We could pass a law that non-therapeutic medical treatment— treatment aimed not at the health of that person but at keeping the person alive for some external purpose—is illegal," Neuhaus suggested. He addressed Imber on this point: "That could be done, couldn't it? If that prohibition were firmly established in the law and in medical practice, then in principle we wouldn't have opened up the floodgates, would we—just because we allowed transplants in some circumstances?"

"I suppose my concern would be the definition of 'therapeutic,' " Imber answered. "That's something that seems to me to have changed over time. The very fact that medical schools can acquire cadavers that haven't been claimed is already an indication that the state permits certain kinds of activity that no other groups besides medical groups have attempted to presume. A system has already been established in our culture for a very long time which says that these kinds of decisions are acceptable in certain ranges. And now we've refined this system, and it will continue to be refined. The decisions we make will look, at some levels, very sophisticated, but maybe at some moment along the way someone will say, 'This isn't appropriate.' But I can't see how the line is so simply drawn. I can't see how philosophizing, thinking our way clearly to the problem, is the way the problem will be solved. Perhaps that's my bias as a sociologist."

"I wonder if defining what is and what isn't therapeutic treatment is so difficult," Neuhaus said.

"It's not difficult—it's impossible," responded Nathanson, to general laughter. "I would remind you that long before the abortion laws fell, we had therapeutic-abortion committees, and already then we could do anything. That word 'therapeutic' is so stretchable."

"But what we're talking about is that whatever treatment or technique is employed has to be aimed reasonably at healing the patient's disease or alleviating the patient's pain," Neuhaus persisted.

"Yes, but there's psychological alleviation, which immediately opens the door to anything," Nathanson replied.

The group remained unconvinced by Neuhaus's proposal.

Richard Stith insisted that no harm is done by keeping the anencephalic child alive for another week to use his or her organs. He suggested that the group opposed Nathanson's vision of bodies retained for spare parts because that vision involves serially maiming living persons, and there's something repulsive about that. "But it's not the mere fact of being kept alive that's repulsive. I don't see anything undignified, wrong, or harmful in prolonged life, assuming that no pain or indignity like maiming or something similar is involved. To say that the person is being harmed by being kept alive seems to me to be accepting the argument of the opposition."

At this point Thomas Oden said, "In my billfold, along with

my driver's license, I have an organ-donor authorization card, countersigned by my wife, which says that upon my death I wish to donate, if needed, the following organs or parts for the purpose of transplantation, therapy, medical research, or education. I've designated 'all or any body organs.' At the time I made that choice, I made it with what I understood to be a moral end in view, with the idea of helping someone who needed my body parts. I still think that's a legitimate objective, and its administration needs to be continued, but without some of the ghoulish conclusions we've been drawing. I don't mind a distribution system, a storage system, a rational means of finding proper recipients of my organs. In this fantasy of a 'spare-parts body shop,' I think we haven't distinguished adequately between a dead person and living body parts. In your comments, Pastor Neuhaus, you used the phrase 'carving up people for parts.' That's a way of caricaturing. No person is being carved up. What we're talking about is the therapeutic use, and perhaps other uses, of tissue."

Neuhaus reminded Oden that "these persons are being kept in functioning order."

"No, they're not persons," Oden replied. "This is where the fundamental anthropological question has been neglected in this afternoon's discussion. The human person is this paradoxical interfacing of body and soul. That's the simplest form of Christian anthropology there is. There is no human person if there is no living body—a body including life, because what soul means is life. A body without a soul is a corpse. But a living body embraces that interface. I want that life protected until the body is dead, and I'm willing to accept the brain-death criterion for the definition of death. At this point our task is to try to see how that interface in classical Christian anthropology impacts on issues like the use of fetal tissue, euthanasia, and other related questions."

How to Slow, Stop, or Reverse the Process

As the end of the conference was approaching, Neuhaus suggested that the final period be used to try to pull some things together. "Let me suggest," he said, "that we address this question: If the return of eugenics is a reality; if there is an assertion of human control through technology and coercive power over the human condition in an attempt to remove as much as possible the inconveniences and contingencies and irrationalities of human existence; if what we're witnessing is something like a slippery slope or something like what Bob Destro talked about this morning, an inexorable legal and lethal logic driven by attitudes—if this is all true, *can anything slow it or stop it or reverse it? And, if so, what factors—legal, cultural, religious, technological—can do it?"*

Oden responded to these questions: "Is the slope inexorable? I don't think it is. I don't agree with Bob Destro that the march of *stare decisis* is inevitable or irreversible. Because if we answer that question 'yes,' not only have we forgotten about all the other reverses that have occurred in history, but also we've tentatively given up our own freedom.

"Your second question—whether it's possible to slow, stop, or reverse the trend—is a serious political question. I think the judiciary and its moral sensitivities are very important in answering that question, more important than the legislative processes and all the preachers and all the social-change agents. So somehow our moral reflection and all the energies that go into a discussion like this need to address the judiciary. I don't know how that can happen. But I do have a small niche in the world. It happens to be the education of persons for the ministry, and I do think there's something to be done here.

"I do think, however, that the last part of our discussion— which was among the most interesting of our four days together but did not, regrettably, directly address Paul's brilliant paper— did to some degree misidentify the slippery slope. The slippery slope that really bothers me is the slope on which we move from the privacy principle, to *Roe v. Wade*, to Karen Ann Quinlan, to death by starvation and dehydration. It's the slope that Bob Destro documents, and I don't think the slide down it is irreversible, but I do think it's a very serious problem."

Noel Reynolds suggested that perhaps the most important task is providing guidance for ordinary people. "I'm impressed that with as much agreement as there is in this room, these problems seem to be insoluble using reason and philosophy," he said. "The tools that we have are very good for helping us clarify implications of ideas that we point out to each other, but this is an abstract exercise in reason. Appointing the right judges can be a very influential thing for particular ends. But in the long run, what counts is for the voters to have the right views—or at least the best sensitivities—on these matters. The kind of complicated distinctions we generate around this table are very unhelpful to people confronted with real-life problems. What they need to know about a particular problem is how their mother and father handled it, and how Uncle So-and-so handled it, and how they should handle it.

"I see two things that can help. One is sensitivities, which can be taught through religious instruction. To the extent that the churches are able to teach their people, that will be a help to ordinary people. For those who don't have this kind of resource, we have a much more pessimistic outlook. Then there's a second thing—and this is where I come to Paul Vitz. I see Paul setting up an argument, a plea, for some kind of divine guidance in these matters, and it seems to me that when, finally, the individual faces a particular situation, and there's a right thing or a wrong thing to do, he or she really does need some kind of divine guidance. When we talk about religious experience, we need to remember that it's usually vitally important for the individual to know that he or she is doing the right thing, to have some kind of divine reassurance in that situation."

"That's undoubtedly true, Noel," said Neuhaus. "Most people aren't going to think these issues through with a great deal of sophistication and with all of the nuances and kinds of arguments that we have engaged. And we'd like to think that most people act upon the understandings of right and wrong and good and evil which their sustaining communities believe are the will of God. But if we're talking about changing law and medical practice and public policy, we're making public arguments. And given the polity of the United States, it doesn't help to say, for example, 'I've gotten some divine guidance about what can and cannot be done with respect to anencephalic children.' And I think that's really the focus of the question 'Can we slow, stop, or reverse what's happening?'

"The other question is, How do we create communities that have courage and confidence in a divinely revealed will of God and thus can guide and sustain their members in the decisions they daily make? Without such communities, public discourse is surely useless. But how do we get at the question of the public discourse, the decision-making, the disputes within the medical profession about what can and cannot be done?"

Reynolds agreed: "I don't discount the importance of participating vigorously in those kinds of discussions and trying to shape that discourse and move it in certain directions. I feel we're obligated to participate, and this conference is a move in that direction, which I applaud. But I think that maybe I'm less optimistic than others in this discussion about the potential of that kind of thing."

Bernard Nathanson asked Neuhaus if he had said that at some point in history the slide down the technological slippery slope had been stopped. Neuhaus said he couldn't think of such a point, except possibly the point at which chemical/biological warfare had been put on hold. Hauerwas noted that Japan had stopped producing gunpowder for a century—a momentary pause.

Neuhaus then suggested that technology had at least been restrained and regulated. "I think that some things, carried to certain limits, start setting off alarms," he continued. "It seems to me that one modest element of hope is the 'Claude Pepper factor.' For obvious reasons, the aging are a potent factor with regard to euthanasia. Steve Post pointed out that in medical ethics, termination of human life isn't listed among the harms that can be done. Terminating life is seen not as a harm but as a good that can be done for a person. But that's cultural-linguistic-ideational sleight of hand. It doesn't seem to me that there's anything inevitable about it."

Neuhaus went on to add, however, that he isn't sanguine about the possibility of slowing, stopping, or reversing the course we are on. "One of the reasons," he said, "is our churches: they are the least hopeful communities for developing a new kind of language of caring and responsibility and human dignity. For the most part they're headed pell-mell down the hill in the Gadarene sweepstakes, which is their notion of being 'prophetic'—being in advance of what the new class has just recently come up with.

"Peter Berger says that when I die, my tombstone will say, 'Richard John Neuhaus, 1936 to whatever date.' Underneath that it will say, 'We're going to turn this around.' But that's not really true. I'm not at all sure we're going to turn this one around," he concluded.

Edward Gaffney, responding to Pastor Neuhaus's apparent pessimism at this point, said, "Neuhaus is a very hopeful and sensible person. When he talks the truth about the self-deception of the churches, it's painful but a part of the reality we have to struggle with. But there's another side to Neuhaus that gives an answer along the lines Tom Oden has suggested. To answer in a resoundingly negative way the question about whether we can turn things around is to ignore other historical reversals and to deny our own human freedom—or perhaps more to the point, to deny the transcendent power of God's grace coming before our freedom, to which Stanley Hauerwas made such emphatic reference. Maybe we're in a total mess at this point in our history, and all we can conclude with at the end of this kind of conference is our shame and our powerlessness over our culture. That might put us in a better place to be more energetic and more responsible to that very culture. I don't think it's inevitable that the march will go on inexorably—I don't think that's what Bob Destro is saying. Destro made a crucial distinction between interpretation that's grounded in the constitutional order and the common-law heritage that has about it a more humble, more self-correcting mechanism. We have to locate these questions not exclusively in the judiciary but in the totality of our instruments for shaping public power, including the mediating institutions. And we get back to Brigitte's point. Maybe we now have this overarching new class that is us, but there's a way in which regulation and control and criticism—which this class can provide—have got to be at least part of the hope that we carry from this meeting. Ultimately I'm left with Noel's sense of a kind of powerlessness, but I don't take that to be grounds for despair. I take that to be grounds for a newfound hope."

The Political Project

At this juncture Jean Elshtain pointed to certain practical possibilities with regard to the political project. "Surrogacy is one; the buying and selling of fetal parts or tissue and the patenting and marketing of new life forms are others. Commercialization must become a political question; it has to be put on the political agenda. We can't just leave it up to the judiciary as particular cases come to the fore. We must say that there are some areas where the market simply cannot be allowed to run. If we don't do that, if we permit licensing and regulation, we will have legitimated these practices, and that in turn will help to provide a wider context within which ordinary people will make their decisions about these practices. It will push people in a direction in which we don't want to see them go. It will establish a context in which a woman might be seen as selfish if she doesn't sell off every part of her husband's body. This strikes me as a possibly viable political project—to get this into the political arena and to fight commercialization."

"In some of these things," Neuhaus said, "I'm sure we would agree that commercialization is not *the* driving force. But in all of them commercialization is *a* force, and to stop that would be to help stop these things."

Gilbert Meilaender added, "Whatever kinds of strategies one might latch onto, there's a sense in which there's a fundamental religious question at stake here. Because if God's really not around to blame for certain things anymore, or to shoulder the responsibility for bringing good out of evil in some sense, and if we remain morally serious about these evils and sufferings, then somebody must take responsibility for them. For that, we look like the best candidates. That shouldn't detract from any of the various possible strategic approaches one might think of, but I do believe there's a religious issue at stake here that can't be easily avoided or ignored.

"Second, I do have considerable confidence in the willingness of whoever these 'ordinary people' are to think and talk about these issues. We haven't found the right way to do it yet. A seven-minute segment on the *Today* show doesn't do it; we need ongoing conversation over a period of time. But I do think a lot of people actually crave it."

Paul Stallsworth, a United Methodist minister, added that "a lot of ordinary people would say, 'There are certain people in our society who are trying to play God.' Unfortunately, we Christians don't know how to speak intelligibly in public about the Creator or about the limitations of the creature. Given this situation, it isn't surprising that people take the reins of being creator and create on their own. So I'd like to follow up on what Gil Meilaender said and suggest that our churches once again learn to speak about the God who creates and sustains. That conversation just might spill over into the larger society."

In summarizing at the end of the discussion of his paper, Paul Vitz said that he raises the issue of the lost mystery of God with the academic community. He sees a possible opening there because this community is a critical filter to the society at large. "I'm not in this because I hope to win. The purpose isn't to succeed but to keep the covenant, to be loyal. It's a grace to be asked to fight in this way in this kind of thing. The outcome is in the hands of God rather directly. I wouldn't say Richard Neuhaus is going to turn it around. But whether it's turned around or not seems to me not to be the essence of the joy of the thing."

"This conference has been a good thing," Vitz went on to say. "There are a lot of good people out there who would like to talk about these things. We're not going to count on the churches, because the churches are now part of the problem, not part of the solution. But there are people who want to hear about what we're discussing—not so much the intellectual answers but the future that we're painting, the slippery slope. If there were a way to get the topics we discussed into the public arena, I think there would be a serious outcry. A lot of intelligent and energetic members of the public would be opposed to some of the implications that have been discussed here."

At this point Neuhaus drew the conference to a conclusion. "We *are* going to see it discussed as you suggest," he said. "The alarm buttons are going to be pressed again and again; we can see where the alarm buttons and the trip wires are. There's going to be a lively, intense, confused, cacophonous debate in our society. It will be larger, I expect, than any of the debates over what most people would point to as the greatest trauma of the last forty years: the possibility of nuclear warfare. What we've been talking about comes much closer to home. It's not a matter of speculating about what might happen; it's a matter of looking at what *is* being done."

Neuhaus then referred to Edward Gaffney's perception of his own pessimism. "I firmly believe that this is God's world, that the world is in God's hands, and that this is a God whose purposes will be fulfilled and have already been fulfilled in Easter. I am persuaded that neither life nor death, neither things in heaven nor things on earth can separate not simply me but us, and the causes of which we are a part, from the saving love of God in Jesus Christ. That, I think, is the Christian proposition. Until Jesus returns again in glory—despite the mysterious ways in which there is much bloodshed and internal contradiction and terror—

the promise is kept. We walk by faith and not by sight. For believing Jews, too, the messianic hope relativizes every existing and proposed and imaginable terror, including all the terrors we've been addressing in this conference.

"Ed noted what I've written about democracy, the liberal democratic tradition and its merits, and why we ought not only to affirm it but also to recommend it to others. All of that I do want to affirm. I think it's the best, most honorable, most humane, most filled-with-possibilities answer to the classic political question: How ought we in the United States to order our lives together? However, I also soberly believe that it is now suffering a severe case of overload. It wasn't wired to take on the kinds of questions now being put into the public square and requiring answers. *Roe v. Wade* is the single clearest, most irrefutable demonstration of this. Questions of great moral moment are being forced into the polity, requiring political decisions for which the polity was never prepared.

"But we haven't the right to despair; it's a matter of duty that we hope. As T. S. Eliot said in those marvelous lines from *Four Quartets,* 'For us, there is only the trying. The rest is not our business.' I take that as a statement not of resignation but of triumphant hope. The result of it all, the conclusion, is God's business. He started the whole thing, and somehow, if Jesus is right, he's going to pull it off."

Participants

in the April Conference and/or the May Conference

Myriam Puig Abuli
Pediatrician
Caracas, Venezuela

Hadley Arkes
Department of Political Science
Amherst College

Brigitte Berger
Department of Sociology
Boston University

James T. Burtchaell, C.S.C.
Department of Theology
University of Notre Dame

Sidney Callahan
Department of Psychology
Mercy College

Francis Canavan, S.J.
Jesuit Community
Fordham University

Robert A. Destro
The Columbus School of Law
The Catholic University of
    America

Jean Bethke Elshtain
Department of Political Science
Vanderbilt University

Hillel Fradkin
The Lynde and Harry Bradley
    Foundation

Edward McGlynn Gaffney, Jr.
Loyola Law School
Los Angeles, CA

Stanley Harakas
Holy Cross School of Theology

Stanley Hauerwas
The Divinity School
Duke University

Richard G. Hutcheson, Jr.
Vienna, Virginia

Jonathan B. Imber
Department of Sociology
Wellesley College

Christopher Lasch
Department of History
University of Rochester

Gilbert Meilaender
Department of Religion
Oberlin College

Kenneth A. Myers
Powhatan, Virginia

Bernard N. Nathanson
Bernadell, Inc.—Educational
    Works on Bio-ethical Issues
New York City

Richard John Neuhaus
New York City

David Novak
Department of Religious Studies
University of Virginia

Thomas C. Oden
Theological School
Drew University

Stephen Post
Center for Biomedical Ethics
School of Medicine
Case Western Reserve University

Noel B. Reynolds
Department of Political Science
Brigham Young University

Barry Schwartz
Department of Psychology
Swarthmore College

Fred Sommers
Department of Philosophy
Brandeis University

Paul T. Stallsworth
New York City

Margaret Steinfels
Commonweal

Richard Stith
School of Law
Valparaiso University

Paul C. Vitz
Department of Psychology
New York University